the Hungry Scientist

HANDBOOK

the Hungry Scientist
HANDBOOK:

[
Electric Birthday Cakes, Edible Origami,

and Other DIY Projects for

Techies, Tinkerers, and Foodies
]

Patrick Buckley and Lily Binns

COLLINS LIVING
An Imprint of HarperCollins Publishers

Every effort has been made to ensure that all the information contained in this book is accurate. This book is not intended to replace manufacturers' instructions in the use of products—always follow their safety guidelines. Also, manufacturers constantly change their products' composition and other specifications and the limitations upon their uses. As a result, some of the projects may not work as described and may adversely affect or damage equipment used. The authors and publisher disclaim any liability for any personal or property injury, loss, or damage caused or sustained as a consequence of the use and application of the contents of this book.

The projects in this book involve the use of tools, resources, and materials that can be dangerous and require training, supervision, and practice in order to be used safely, and the projects themselves may be dangerous if not done exactly as directed. Your safety is your responsibility, including the proper use of equipment and safety gear and determining whether you have the adequate skill and experience required for a project. If at any time you do not fully understand either the instructions for a project or the dangers involved in handling some of the required tools or materials, you should immediately discontinue the project and consult with someone who does. Please note that some of the illustrative photographs do not depict safety precautions or equipment, in order to show the project steps more clearly.

This book is not intended for children. If children attempt any of these projects, they should always be supervised by adults who have experience with the tools, resources, and materials required.

THE HUNGRY SCIENTIST HANDBOOK. Copyright © 2008 by PATRICK BUCKLEY AND LILY BINNS. All rights reserved. Printed in the United States of America. No part of this book may be used or reproduced in any manner whatsoever without written permission except in the case of brief quotations embodied in critical articles and reviews. For information, address HarperCollins Publishers, 10 East 53rd Street, New York, NY 10022.

HarperCollins books may be purchased for educational, business, or sales promotional use. For information, please write: Special Markets Department, HarperCollins Publishers, 10 East 53rd Street, New York, NY 10022.

FIRST EDITION

Designed by Jaime Putorti

Library of Congress Cataloging-in-Publication Data

Buckley, Patrick.
    The hungry scientist handbook : electric birthday cakes, edible origami, and other DIY projects for techies, tinkerers, and foodies / by Patrick Buckley and Lily Binns. — 1st ed.
        p.   cm.
    Includes index.
    ISBN 978-0-06-123868-0
    1. Technology—Experiments.   2. Science—Experiments.
3. Food—Experiments.   I. Binns, Lily.   II. Title.

    T65.3.B83 2008
    643'.3—dc22   2008005687

08   09   10   11   12   WBC/QW   10   9   8   7   6   5   4   3   2   1

T hank you, Hungry Scientist contributors. You are some of the most brilliant people with whom we've ever had the opportunity to break bread and set things on fire. And thank you, Anna Bliss, Toni Sciarra, and Anne Cole, our wonderful editors, for bringing this strange book to life.

The Hungry Scientist Handbook is dedicated to our mothers, Judith Regan and Nancy Binns.

contents

Here's the part we'll get sued for if we don't include: There are some serious hazards to carrying out many of these projects at home. For example, were you to wear edible caramel lingerie (chapter 1) while flaming up a beer-can stove (chapter 8), you might end up with a bikini wax you didn't plan for. Needless to say, be—as they say in Oakland—*helluv* careful while wielding blowtorches and tossing about frozen carbon dioxide. We'd like to be responsible for your fun but not your shattered fingers!

Serving **Hungry** Science

This is a handbook for extreme comestible creativity. After all, scientists are the ultimate cooks. Techies, tinkerers, and do-it-yourself home improvers tend to figure out how to cook the same way they solve other problems: Whether they're creating a meringue pie or a megaphone, they are inventive, resourceful, daring, and determined. When Hungry Scientists gather in the kitchen to collaborate on contraptions and cuisine, no tools, equipment, or ingredients are off-limits.

This book is a guide to turning your kitchen into a lab and a workshop. It includes instructions for making some of the most delectable inventions we've come across. We know, however, that inventors have no qualms about departing from rules. We hope this book inspires you to concoct and construct even bigger and bubblier creations. It'll be hard to top the likes of these projects (witness the modular pecan pie, chapter 18), but we suspect you've got the brains to do it.

The Hungry Scientist Handbook started with a dinner party that the two of us threw a few years ago. Patrick, a mechanical engineer working in the medical department at Lawrence Livermore National Laboratories in California's Bay Area, invited twenty friends and colleagues into his small bachelor pad to eat a sumptuous meal. Kaori Kuribashi, a scientist visiting from Japan who was developing prototypes of shape-memory alloy devices that are used to repair collapsed arteries, inspired the occasion. At the time, Patrick was working on similar devices made from shape-memory polymers at his lab. The goal of the evening was to toss around their ideas with a group of like-minded scientists and gadgeteers.

Lily—one of Patrick's oldest and closest friends, an avid cook, and, at the time, a cookbook editor—engineered a dinner to sharpen the guests' minds and raise their spirits. And what a meal she invented. By the end of the evening, as the last of the succulent roasted fowl (see chapter 11 for a creative roasting contrivance) was picked from its platter and a postprandial satiation quieted the brainiac discussion, the Hungry Scientist Society had been born.

After the success of its first gathering, the Hungry Scientist crowd grew and multiplied. Joined by local foodie members of the burgeoning DIY Web site www. instructables.com, we had more dinner parties to discuss inventions, where the food started to become the projects in question. We noticed our friends' unusually strong interest in not only teaching themselves how to cook but also in cooking outrageously ambitious dishes. Sometimes they were imaginatively resourceful, creating, for instance, whole meals for less than a dollar. Often they took making dishes from scratch to a whole new level (see bread made from wild yeast, chapter 15, or wine from pomegranates, chapter 16). All the projects exhibited a certain set of common chromosomes: wild creativity, humor and playfulness, and serious smarts.

This book is a collection of the best recipes for gadgets and schematics for food that we've encountered so far. Whether they're edible or not, we think all of the projects are irresistible. Trust us. We're geeks, and we're food people.

This is our manifesto: Gadgeteering and gastronomy are about having fun. They are community crafts based on both ancient principles and new technologies. To people like us, who have chronic attention deficit and live frenetic lives, these are all important things, and we think they deserve celebration. We put out an invitation through that milk-bottle megaphone (chapter 10) to bring more ideas to the table, and the projects that follow are what came in. The kitchen is infinite, friends. This book is just the beginning.

Join us at the table and share your own projects at

www.hungryscientist.com

chapter 1

Edible Undies

* *Lace-Up caramel lingerie*

We had purely scientific intentions when we went into development for **an edible lingerie prototype**. Seriously—haven't *you* ever wondered **how a chocolate thong works**? We asked our fashionista-inventor friend **Jenna Phillips** if she knew how to go about designing undergarments suitable for salivation and mastication. Needless to say, we were very **pleased by her invention**.

Caramel candy is formed by caramelizing sugar, or heating it until the sucrose molecules break down into different compounds, which become darker in color and toastier in flavor the longer they're cooked. The more you heat, let cool, and reheat the caramel, the stronger it will become—though you probably shouldn't count on this lace holding up all night. Then again, it's not really meant to last.

Equipment

10-inch plastic-coated icing bag

A variety of small-opening decorating tips for icing

Wax paper or silicone mats

Lingerie, to use as template

Felt-tip pen

Ingredients

1 stick butter

1⅛ cups light brown sugar

½ cup dark Karo syrup

½ cup condensed milk

¾ tsp vanilla extract

[MONSIEUR **MAILLARD**]

Caramel's flavor is deepened by adding milk to the mix. When carbohydrates (sugar) are heated with the amino acids in protein molecules (milk), they undergo what's known as the Maillard (my-ARE) reaction. It is typically responsible for the browning that occurs in savory foods such as roasted meat, bread crusts, dark beer, coffee beans, and chocolate, giving them their toasted flavors and aromas. We wager that it contributes to the complex, deeply sensual character of this confection.

So, we dedicate this project to the French physician and biochemist Louis-Camille Maillard, discoverer of the browning reactions that bear his name. Among other topics, he researched the chemical origins of the different ways foods taste when they're cooked. Were Maillard still around today, we'd love to engage him in our edible lingerie research.

[TIPS FOR **MAKING CARAMEL**]

Caramel is best made in a cool, dry atmosphere. Sugar is hygroscopic, which means it pulls moisture from the air, causing the caramel cooking time to increase and take longer to set up. If it's made on a rainy or humid day, the concoction will absorb so much water that it'll turn to syrup rather than viscous caramel.

1 To create a template, stretch a piece of lingerie out over a sheet of wax paper and trace its outline with the felt-tip pen. Fold the paper in half to check that its shape is symmetrical. Attach the smallest tip to the nonstick icing bag and set aside.

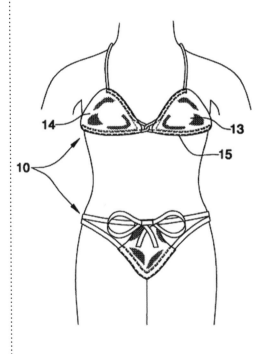

2 To concoct the caramel, combine the butter and sugar in a double boiler or a heavy-bottomed pan. Add the Karo syrup and evaporated milk. Stirring constantly, cook over low heat until the mixture starts to thicken. Test the thickness by spooning threads of it onto a piece of wax paper. After it cools for 5 minutes, the caramel should be pliable, but not too liquid; runny enough to pass through the small tip of the icing bag but thick enough to hold its shape.

3 Holding the icing bag in a dish towel to prevent it from burning your hands, spoon about ½ cup of hot caramel into the bag. Fold down the top of the bag to close it, and, squeezing from the top down, drizzle the caramel in a lacy pattern within your template.

4 The caramel will quickly start to cool in the bag. When it becomes too viscous to pass through the tip, scrape it back out with a spoon into the pot on the stove, and reheat it until it's runny again. You will need to reheat the caramel several times. This may seem like a pain, but the reheating becomes important later in the process of creating the lingerie.

5 To attach the different pieces of lacework once the templates have been sufficiently filled out, you will need stronger pieces of caramel to act as connectors. Scrape the caramel that has been reheated several times into the icing bag. Attach a larger decorating tip and squeeze out three long ropes of caramel. Lay them next to one another on the wax paper and braid them.

6 As the braids cool and stiffen, attach them to the top of the lace with dabs of hot caramel. Keep a few longer strips to go around the back. (You will attach these pieces to one another when putting the lingerie on a person.)

7 Now that the lingerie is close to completion, set it in your freezer for 5 to 10 minutes to let it cool and harden. When you remove it from the freezer, it should not feel sticky, and will easily peel off the wax paper.

8 Place the entire piece of lingerie on a mannequin or willing human subject.* With small dabs of hot caramel, attach the final connectors. As the lingerie warms to body temperature, it will become pliable, and very sticky. If sticky isn't your style, cook the caramel longer at the beginning. The more it has been cooked, the harder it is, and the slower it melts. Either way, the result is delicious.

See the Appendix for a source for more information on candy making and supplies.

*
No lingerie models were harmed during the undertaking of this experiment.

3a

3b

8a

8b

8c

chapter 2

Delectable **Diodes**

* *Brighten lollipops with LEDs*

Our man **Dan Goldwater**, aka Señor LED, told us he wanted to make an edge-lit cake, by borrowing the principles from those **high-tech flat-screen television displays** that funnel photons through their plastic edges, but using instead a caramel glaze. As he experimented with LEDs and their lighting properties in caramel, a **puddle of caramel** formed around a stick and a single LED light, which together looked a whole lot like a lollipop. The spectacularly lit LED lolli was born.

Traditional hard-candy recipes call for corn syrup and cream of tartar, which disrupt the formation of sugar crystals and lead to clear lollipops and lozenges. When lollipops are made solely with sugar and water, however, large sugar crystals form as the mixture cools. The multisided crystals refract the light from the LED and scatter it throughout the lolli, increasing its beautiful glow. It's your choice whether to go with clear lollipops or ones that refract.

Equipment

1 empty plastic container, such as a yogurt cup or other mold form

13-V LED

2 (6- to 8-inch) electrical wires

1 empty plastic ballpoint pen barrel, preferably clear

Soldering iron

Solder

Electrical tape or shrink tubing

Packing material, such as toilet paper or orange rind

1 (CR2032) 3-V lithium coin battery

Candy thermometer (optional)

Ingredients

1 cup sugar

¼ cup corn syrup (optional)

½ cup water

¼ tsp cream of tartar (optional)

¼ to 1 tsp flavoring (optional)

[LED **LOVE**]

 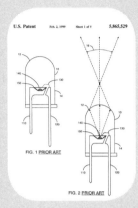

Think of LEDs, or light-emitting diodes, as layer cakes with half-micron-thick layers. The diode's layer is made of wafers of raw material, sliced from an ingot of semiconductor materials, such as gallium arsenide or gallium phosphide. The slices are doused with a variety of chemicals such as zinc, silicon, or nitrogen (think of these as the flavoring of your cake)

The unique layering of materials in an LED determines what is known as the electron band gap. This band gap is the distance that electrons must jump between layers of your LED cake. Much as the filling in a layer cake determines the flavor of the dessert, the electron band gap determines the color of light emitted by an LED. For example, when phosphide—think of it as lemon extract—is added to a diode's wafer, it changes the electrons' band gap, and thus changes its wavelength, which is turn determines the color of the light that the respective diode puts out. In the case of phosphide, it's yellow.

So, LEDs are electroluminescent. They generate light by electronic excitation rather than by heat generation, and their energy flows in only one direction. They are highly efficient, requiring little energy and emitting almost no heat. The more electrons that pass across the boundary between layers, the brighter the little light.

1   To prepare a lollipop mold, cut off the bottom of the yogurt container about ½ inch from its base. Cut a small notch in the side of the mold slightly smaller than the size of the pen's barrel.

2   Thread two 6- to 8-inch-long electrical wires through the empty pen barrel and solder them to the LED's wire legs. The wires need to be long enough to extend out the opposite end of the pen. Using tape or shrink tubing, cover at least one of the bare metal connections so that the two legs don't touch inside the barrel and cause an electrical short.

3   Place a small bit of packing material around the LED legs to seal the end of the tube. Pull the wires gently until the LED and seal are flush to the end of the tube.

4   To make the lollipop, mix the sugar and water (and other optional ingredients, if wanted) in a small Pyrex, enamel, or nonstick pot. Put that pot in a medium heavy-bottomed pot and fill the bottom pot with an inch of water, place over medium heat. Bring the mixture in the small pot to a boil. Allow it to boil until the liquid reaches 300°F, or the hard-crack stage. If you don't have a thermometer, test the temperature of the mixture by pouring a few drops of the liquid into cold water. When it's ready, it will form hard, brittle threads that break when bent. (The hard-crack stage lasts only for a few seconds, beyond which the sugar will darken as its molecules begin to break down, so watch the liquid closely to make sure it doesn't burn.) As soon as the hard-crack stage is reached, remove the mixture from the heat and set it aside for a few minutes to cool, until it's the consistency of honey.

5   Put a small amount of ice water on a tray or plate. Swab the mold with a small amount of vegetable oil to prevent sticking. Put the mold into the ice water, and place the pen lights into the molds.

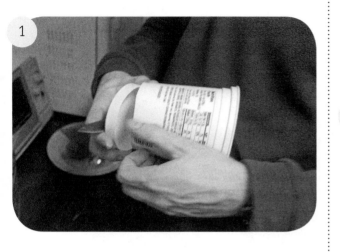

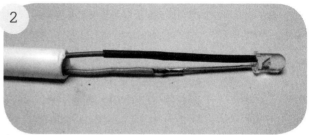

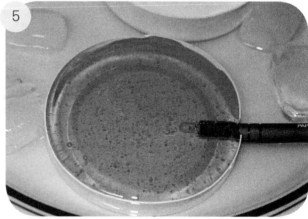

6 Carefully pour the mixture into the mold. Make sure that water does not leak around the tube and dissolve the lollipop. Allow the lollipop to harden (between 1 and 2 hours). When it is completely hard, pop the lollipop out of the mold. Attach the battery to the wires with the electrical tape. (If the diode doesn't light up, switch the battery around.) Lick away.

* *See the Appendix for more information on LED voltage and wiring instructions.*

chapter 3

Pumpkin Pin-Up

* *Make a pinhole camera out of (almost) anything in the kitchen*

Photography can be **an expensive habit**, but standard kitchen equipment and a pumpkin will supply **everything you need** to make a simple camera (and a tasty dessert). We asked our foxy photographer friend **Kate Gusmano** to show us how.

Pinhole cameras are fascinatingly simple inventions capable of producing sharp, unique photographic images without the use of a lens. Employing the basic principles of photography at its most low-tech, a pinhole camera is an extension of the camera obscura—a darkened room with a small opening to admit light. The camera obscura has long been used for scientific and artistic experimentation, from astronomical observation to illustration to theater.

Kitchens are full of things that make great cameras: any light-tight container will work. In the perishable category, look for items with a thick rind or shell that can be hollowed out. For cameras with a longer shelf life, oatmeal boxes and cookie tins are good choices. When possible, paint the inside black to prevent light from bouncing around and possibly affecting the film.

[DIY DARKROOM BASICS]

Pinhole photography requires access to a darkroom, but anyone can make a darkroom at home. One can be constructed in any well-ventilated space where light can be blocked out. To rig a DIY darkroom, set up three plastic trays: the first with developer solution, the second with stop (you can use water), and the third with fix solution. You'll also need a safelight, tongs to handle the paper, and a place to hang the negatives while they are drying. Instructions for developing the paper will vary depending on the types of chemical solutions used and their concentrations; follow the manufacturer's suggestions. This is all you will need to develop paper negatives. These may be scanned to produce positives. They may also be contact-printed in a darkroom, requiring the use of a photographic enlarger. Unless you own one or plan on investing in an enlarger, a commercial darkroom may be the only option for making your prints.

[JACK-O'-LANTERNS
AND PUMPKIN PIES]

If you use a pumpkin to make your pinhole camera, you can of course transform it into a traditional jack-o'-lantern. To keep it light-tight, paint on the other features instead of carving them. If you've used an edible variety, such as a squash or a sugar pie pumpkin, you can transform it, yet again, into a pie. If you are planning to bake your creation, keep it clean and use it within a day or two of carving. When you are ready to make the pie, clean the pumpkin and cut it into chunks. Bake the pieces skin-side down at 375° until soft. Once they have cooled, peel off the skin (discard it) and puree the pulp until smooth. Use the puree in place of canned pumpkin in your favorite recipe.

Equipment

1 pumpkin

Pumpkin carving tools

1 aluminum pie plate

1 size-10 sewing needle

Fine-grade sandpaper

Resin-coated photographic paper, preferably matte (to serve as your film)

Scissors

Black mat board or illustration board (optional)

Lightproof black tape (gaffer's tape works well; so does black duct or electrical tape)

Changing bag (optional)

Access to a darkroom and black-and-white printing equipment

1 Slice the pumpkin from top to bottom, beginning about ¾ of the way across the top. Scoop out and discard the seeds and webbing. Using a sharp knife or carving tool, cut a small hole through the front of the larger half of the pumpkin.

2 Cut a piece of the pie plate slightly larger than the hole in the pumpkin. Using a sewing needle, make a hole in the plate, pressing gently so that it doesn't warp. Sand the rough edges.

3 Cover the hole in the pumpkin with the plate, using electrical tape to attach it to the outside. Be sure the tape seals out all light.

4 In a darkroom, trim the photographic paper so that it fits snugly inside the cavity at the back of the pumpkin, opposite the pinhole. You may want to create a film holder using two pieces of black mat board or illustration board glued together so that the photographic paper can slide between them. Cut a window in one board. The opening should be slightly smaller than the paper. A slit may be made in the top of the pumpkin so that the holder can be easily removed (and resealed later to prevent light leaks).

5 In a darkroom or changing bag, load the paper/film (with the coated side facing the pinhole), one sheet at a time, and tape the pumpkin closed with electrical tape so that no light can enter. Place a piece of tape over the pinhole.

6 Place the pumpkin on a stable surface, preferably outdoors in bright light. Try not to photograph into deep shade or direct light. To make a photograph, peel back the tape covering the pinhole, taking care not to move the pumpkin. Experiment with exposure times. It may take anywhere from seconds in a shallow cavity to several minutes in a deeper cavity, depending on the distance from the pinhole to the paper/film and the amount of light entering. In full sun, for instance, you may want to try an exposure

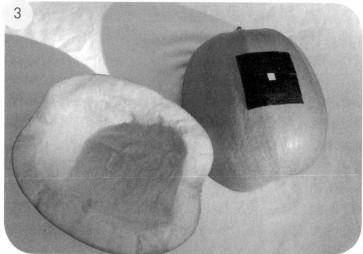

at 10 seconds, 30 seconds, 2 minutes, and 5 minutes. Remove and reload film between exposures in a darkroom or changing bag.

7 Process the paper in a darkroom set up to make black-and-white prints, using manufacturer-recommended times for developer, stop, and fix solutions. Hang the prints to dry using wooden clothespins (mechanical dryers eat small pieces of paper). This will yield a paper negative. A positive print can be easily made once the negative is dry by placing it facedown on another piece of photographic paper under a sheet of glass or in a contact printer and exposing it under an enlarger. Process the positive.

8 Because the exposure times are relatively long, stationary subject matter works best. Moving objects may not register at all on the film. The size of the image will vary depending on the distance between the pinhole and paper. A greater distance requires a longer exposure time and creates a larger image. A shorter distance creates a smaller image and also a wider angle of view (more of the scene is recorded). The sharpness of the image will also vary with the size of the pinhole and the focal-length distance from the pinhole to the paper. See the next page for some combinations that create sharp images:

Needle #	Pinhole Diameter	Focal Length
8	.023"	8"
9	.020"	6.5"
10	.018"	5"
12	.016"	4"
12	.013"	2.5"

* *See the Appendix for a source for photo-developer supplies.*

Party Like **It's 2099**

Light up a birthday cake with LED candles

What could be cooler than **a cake with conductive frosting?** Here at **Hungry Scientist Headquarters**, we set out to make an edible electronic centerpiece: an electron-powered birthday cake with LED candles **good enough to eat**.

Developing an edible circuit was a long and arduous journey. First, we hit the bottle—Goldschlager, to be exact, for the signature gold leaf that floats amidst the alcohol. Though it worked wonders in raising our creative spirits, skimming enough leaf to make a line circuit atop a frosted cake was difficult and far too expensive to recommend. Sobered, we attempted an electrolyte-saturated sports-drink-powder frosting. Unsurprisingly, we found that the frosting's low water content prevented the drink's charged salt ions from freely moving around and carrying electrons. Desperate, we even tried nontoxic electrode pad gel, which did work but was disgusting.

Along the way, we stumbled across the atomically thin edible silver leaf called *varak* that is traditionally used as a garnish for Indian sweets. The idea of applying metal leaf to sugar got us back on the right track, and we finally achieved total success by wrapping *varak* around shoestring licorice—a perfect, edible conducting wire for our circuit.

Consumed in minuscule amounts, silver has been proven edible through the centuries, but we can't recommend eating a ton of it. It's bio-accumulative and stays in the human system for life—which is a cool concept, but we don't want you to become a good conductor of electricity yourself.

[ELEMONICITY]

If you're a purist and want to go organic, use a lemon battery to power your LEDs. The citric acid in lemons acts as an electrolyte, and when combined with electrodes of the right material, such as copper and zinc, the resulting chemical reaction creates free electrons—voilà electricity! Electrodes at either end of a lemon can be created with a galvanized nail and a copper penny (use a pre-1983-vintage U.S. penny; any newer and they're made with copper-coated zinc). An average juicy lemon will generate only about 1 volt of electricity, and since a typical red LED requires at least 2 volts to light up (and some other colors as much as 4½ volts [see page 9 for LED Love]), you'll need a mountain of lemons to power your cake, depending on the age of the birthday girl or boy.

Equipment

1 multimeter

Batteries (9-V or AA or AAA will work, but the type of battery you use depends on the design of your circuit: see steps 1–3)

LEDs of the same color, check required voltages (see Appendix for sources)

1 Electronics breadboard (recommended; can be found at RadioShack)

Ingredients

1 frosted cake

Shoestring licorice (preferably a sticky version, such as Twizzlers Pull-n-Peel)

Edible silver or *varak* leaf (see Appendix for a source)

1 Decide what color and how many LEDs you want to use. Stick to one color and model of LED; it will make the design of your circuit much easier, because all of the LEDs will have the same voltage requirements.

2 Divide the voltage of your battery (Vb) by the voltage requirement of your chosen LED (Vled)—these numbers should be provided by the manufacturers. The outcome, or cluster count, needs to be a whole number. If it is not a whole number, you need to pick a different LED and battery combination. We used a 9-V battery and blue 3-V LEDs: Vb/Vled = 3 = perfect. This is the cluster count we will use in the next steps.

3 Draw your circuit and test your design on an electronics breadboard before you lay it out on your cake: LEDs have a polarity; they will work only when electricity is flowing from their positive to their negative terminals. With this polarity in mind, you have two options for wiring multiple LEDs together: in series (from the negative terminal of one to the positive

terminal of the next), or parallel (from positive terminal to positive terminal and from negative terminal to negative terminal)—see photo 3a. This is where the cluster count becomes useful. The cluster count tells you how many LEDs you need to wire together in a series so that their electrical load will match the voltage of the battery, thus making the LEDs light up but not burn out. These clusters of LEDs are then wired as a group into the rest of your circuit in parallels. This means that the total number of LEDs must be a multiple of your cluster count. So, for the cake we built, the cluster count of three limited us to using a total number of LEDs that was a multiple of three.

4 Roll the licorice in the sheets of silver leaf to completely coat each string: Stretch the licorice a bit or wet it, to make the surface tacky enough for the foil to stick. The sheets of leaf can overlap and be ragged; what matters is that the licorice, especially at its ends, is well covered.

5 Test the conductivity of the wires by touching them at each end with a multimeter set to read ohms and measure the electrical resistance. The resistance should basically be zero.

6 Build the circuit atop your cake: When wiring it is helpful to hook the battery up first and wire an entire cluster of LEDs at once. That way you can see each cluster of LEDs light up as you wire your cake, giving you quick feedback about whether all of your licorice connections are working. If they aren't, clean the ends of the licorice, add more silver if necessary, and reset the circuit atop your cake.

Using your diagram, lay out the wiring for your circuit with your edible licorice wire. To make longer sections of wire, overlap the ends of the edible wire, making sure that frosting does not sneak in between the wires.

You will need to bend the leads of your LEDs to right angles so that they can straddle your two wires (see picture). Be careful to wire the LEDs with the polarity in the right direction. Usually the longer lead is the positive terminal of the LED. You can check this by holding the LED leads

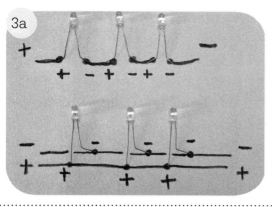

3a

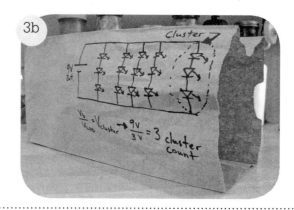

3b

$$\frac{V_b}{V_{LED}} = V_{cluster} \rightarrow \frac{9V}{3V} = 3 \text{ cluster count}$$

3c

4a

4b

5

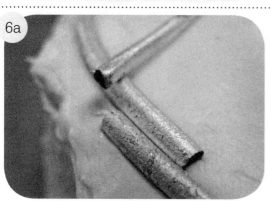

6a

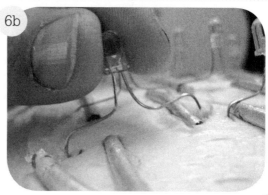

6b

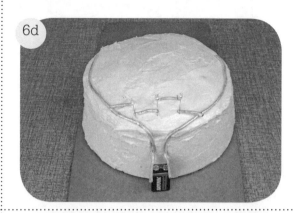

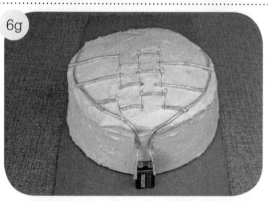

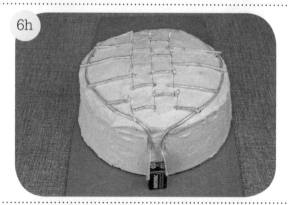

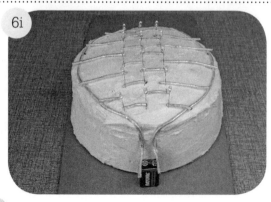

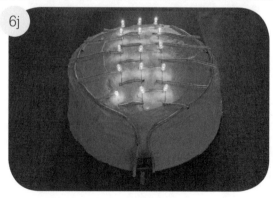

THE HUNGRY SCIENTIST

to the + and - terminals of your battery to see if it lights up; don't hold it there for more than a second as you will burn out your LED. If the LED doesn't light up, flip the leads around.

7 Enjoy your glowing masterpiece. But don't eat the LEDs; they still need to be pulled out of the cake, just like candles. The rest of your circuit is good enough to eat—a tiny bit of and only once a year!

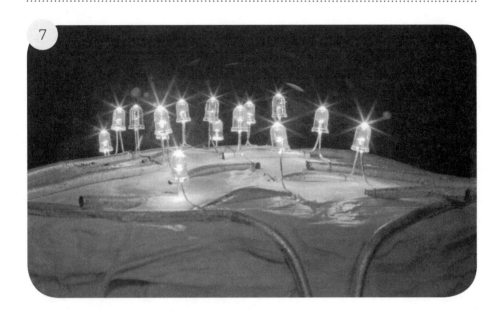

7

chapter 5

Dip 'n' Dots

Assemble a computer chip trivet

Every good geek has a supply of scavenged integrated circuits, some of which may actually be marked "Dead?" **Windell Oskay** and **Lenore Edman** found a way to give them new life. Creating the perfect arrangement of chips is like playing **a game of tangrams or Tetris**—it isn't as easy as it looks! Give those **trendy DIY home-dec hipsters** a run for their money as you turn your old chips into a high-temperature trivet that holds kitchen creations when they come hot to the table.

Equipment

A big pile of integrated circuits in ceramic packages

A ceramic tile, approximately 8 × 8 inches, or large enough to hold a skillet

Adhesive caulk or RTV silicone adhesive, any variety suitable for nonporous materials

Sanded grout, the "just-add-water" type, your choice of color (5 lb is enough for a trivet or two)

Newspaper, newsprint, or a drop cloth

Disposable cups and spoons

Cotton swabs

Optional

A squeegee (strongly recommended)

Rubber gloves, balloons, finger cots, or similar devices to keep your fingers clean

Rubber feet

Margaritas

Tile sealant

[CHIP CHIP **HOORAY**!]

Any integrated circuits will do for this project so long as their packages are made of metal, glass, and/or ceramic. Plastic is out. We used EPROMs with pretty quartz windows, the 68020 and 68881 from a Mac II, and a massive DSP chip, among others. Note that the metal caps on some of the pretty ceramic chips might loosen under extreme heat. If you don't have or can't scrounge enough chips, you can supplement your pile with wire-wound power resistors in aluminum or ceramic ("sandstone") cases, rectangular rectifier bridges, ferrite transformer cores, crystal oscillator cans, chunks of plate glass, and optical prism assemblies removed from DVD players.

1 Arrange your chips on top of the rough side of the tile. Leave a ¼-inch border around the outside, and ⅛ to ¼ inch between adjacent chip edges. If you have chips with significantly different thicknesses, try to arrange the thicker ones symmetrically so that they will support your pots and pans evenly.

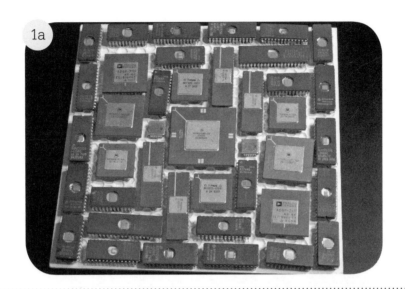

Take care not to leave gaps too large or too small between chips, as shown in 1b.

2   Glue the chips to the tile with the adhesive: Lift one chip at a time and apply a narrow, tall bead of adhesive in its place. Gently place the chip on the bead of glue and press it firmly onto the tile. Use just enough adhesive to partially fill the area under each chip. There needs to be room for grout in the areas between chips because the grout will not stick to caulk. Let the adhesive dry. (This may take up to 48 hours, depending on the type of adhesive you use.)

3   To mix the grout, fill a disposable cup ⅓ of the way with dry grout powder. Add water slowly, stirring thoroughly. Aim for the consistency of slightly runny toothpaste. If it gets too thin, mix in more dry grout.

4   Put the newsprint or a drop cloth down under your tile. Serve the margaritas. Make sure that you're ready to get started, since once you begin grouting, you're committed to finishing in one sitting, before it hardens. Use a spoon to shovel the grout into the trenches between your chips. The grout should completely fill the trenches all the way down to the tile. The tops of the chips will get quite messy in the process. Don't worry—you'll clean it up in the next few steps. Don't worry about getting grout on the edge of the tile yet either; there's time for that later.

5   Using a squeegee, wetted cotton swabs, fingers protected by rubber gloves or, if you're a guitar player, your unprotected fingers, begin to smooth out the grout. Try to clean all the grout off the tops of your chips and leave a smooth, recessed trough between them. Once you've roughly cleaned the chips, begin to sculpt the border of the trivet by adding a thick bead of grout to the outside edge of the tile to build up a smooth transition between the tile and the chips. Make it pretty. Using cotton swabs, wipe the remaining grout off the tops of the chips. It is much easier to do that while the grout is still wet. Leftover grout dust on the chips can be removed with cotton swabs, but it takes a little more effort. You will also find some loose sand particles on the outer surface of the grout. Wipe the surfaces lightly with your hands or a dry paper towel to remove this debris.

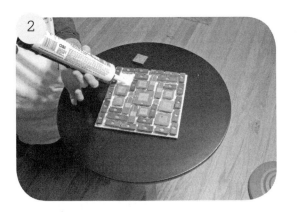

6 You may want to add rubber feet to the bottom of your trivet. Rubber feet will help to keep a trivet from sliding, though the adhesive that holds rubber feet in place may loosen when it gets hot. You may want to paint your new trivet with a grout sealant that waterproofs the grout and gives it a pleasantly slick appearance. Sealant tends not to be designed for high-temperature use, however, so apply one only if your trivet is purely decorative or will be used for cold, wet items, like margaritas.

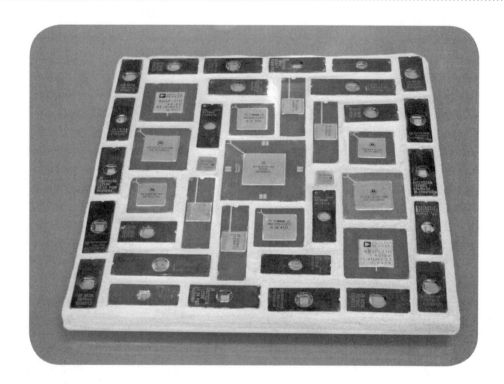

Magnet Madness

Our buddy Christian Brookfield *got crafty and covered a fridge with spreadable cheese, Yoo-Hoo, raisins, candy, and chewing gum.*

To create magnets out of containers of tasty treats, unfold one flap on the top of a box and, using a **hobby knife** delicately make an incision such that the incision will be hidden if you fold the flap down again. Squeeze open the containers and drizzle **hot glue** down the inside wall. Since high temperatures can damage magnets, wait as long as possible before the glue begins to dry. Place the magnet in the pile of hot glue, and push it down hard against the wall. Hold it in place until the glue is set, which should be less than a minute.

To seal up the box, fold the top flap back down, hiding your incision. Apply a drop of hot glue under each of the side flaps. Hold down both of the side flaps until the glue sets. Attach to fridge.

[DID YOU **KNOW**?]

Both Good & Plenty and Nerds contain carmine, a pigment made from crushed insects. Sugar Babies and Good & Plenty both contain confectioners glaze, or resinous glaze, which is derived from the lac insect. People with vegan, vegetarian, Kosher, halal, and other types of non-bug-eating diets: beware!

chapter 6

Bar **None**

* *Concoct super-chilled martinis,*
fizzy lemonade, and über-bubbly root beer

There's nothing quite like a **Hungry Scientist bartender**

winking at you through the smoke billowing from the top of a

dry-ice martini. In fact, **we swooned** when **Windell Oskay**

poured us one. Following suit, **Lenore Edman** fizzed up some

old-school lemonade for us, and **Christian Brookfield** con-

cocted dangerously bubbly root beer—all with a **hunk of dry ice**

fetched from the local market.

When carbon dioxide breaks down from its frozen state (dry ice), it sublimes directly to gas rather than liquid. It's excellent for chilling drinks without watering them down, and it also makes bubbles violently levitate from the bottoms of glasses, which is, well, just trippy to watch.

Many supermarkets carry dry ice, and we've included a source in the Appendix for finding it in your neighborhood. Our local shop sells it for about $1.50 per pound. A 10-pound block should suffice for all of these recipes. Be sure to take a cooler to the store with you for transporting your dry ice home, and use gloves and even goggles when handling. Use common sense! IT'S REALLY COLD! (-109.3°F.) Don't hold the dry ice with your bare hands or do anything else to show off with it!

 CAUTION: *Glass and plastic bottles can break from the stress that extreme temperature changes cause, so don't shake these drinks in irreplaceable containers. Always shake bar shakers away from your face and eyes—and not toward anyone else's face and eyes, either.*

Dry-Ice Martinis

We might be inventors, but we are classicists at heart, and when it comes to martinis, we think there's only one way to make them: with fine gin. (Hendrick's is the best.) Martinis are known for having a half-life of five minutes, since the flecks of H_2O ice that come out of the shaker melt quickly and water down the drink. One way to make martinis extra-cold and preserve them longer is to use frozen CO_2, which can cool the alcohol to a lower temperature than frozen H_2O can.

Ingredients

2½ oz gin

½ oz vermouth, or more or less, to taste

Olives or lemon peel, for garnish

A few small (¼–¾-inch) chunks dry ice

Equipment

Martini glass, chilled

Bar shaker, chilled

Strainer

1 Combine gin and vermouth in the bar shaker. Add a few small chunks of dry ice to the shaker. As soon as you do so, it will start bubbling like crazy, spewing out gas at a terrific rate. You'll have to move quickly; you don't want your martini particularly carbonated, and you don't want it freezing completely solid.

2 The outgassing combined with the chilly temperature makes it difficult—but not impossible—to put the lid on the shaker and shake it.

Better to swirl it. (This is a case in which stirred is much better than shaken.) It will not take long to make the martini sufficiently cold.

3 Pour the part of the martini that's still liquid through the strainer into the glass. Since so much of the martini will have frozen into tiny crystals, the liquid will be translucent, with the consistency of honey. Serve immediately. Serves 1.

Fizzy Lemonade

Our lemonade recipe is for the British version, which is carbonated and makes for an excellent alcoholic mixer. We're currently selling it at our 21-and-over lemonade stand.

Meyer lemons are the best lemons to use for lemonade. They're a dark yellow, thin-skinned, not very acidic hybrid of an orange and a lemon. They're relatively common now in markets, and they smell so sweet that we highly recommend planting a small Meyer lemon tree in a bucket in your house to use for daily fizzy lemonade consumption.

Ingredients	Equipment
½ lemon plus 1 slice for garnish	Tall drinking glass
¼ cup sugar	Knife
1 cup hot water	Cutting board
1-to 2½-inch-wide chunks of dry ice	Measuring cup
	A stirring implement

1 Add ¼ cup sugar, the juice of the half lemon, and 1 cup of hot water to the glass. Hot water makes the dry ice react more vigorously, and the longer the drink takes to cool, the longer the CO_2 will bubble through it, making it all the fizzier. Stir well to dissolve the sugar. Add a chunk of ice. The drink will bubble violently for a few minutes.

2 Once the bubbling dies down, keep a close eye on the smoky bubbles as they pop. Each results in a puff of water vapor above the surface. You can often see perfect little "smoke" rings produced at the surface—called toroidal vortices—that can travel, as far as we've observed, to about a foot above the drink. It's easier to see this when the air is still or when you do this in a large bucket, like the one shown in the root beer recipe below.

3 Once the dry ice has disappeared, your result should be cold, tasty, and slightly fizzy lemonade. Garnish with a lemon slice, and, for optimal effect, a chip of dry ice when serving—but don't eat the ice! Serves 1.

Über-Bubbly Root Beer

Really old-school root beer was made of some combination of the following flavors: sassafrass, vanilla, cherry-tree bark, licorice root, sarsaparilla root, nutmeg, anise, molasses, allspice, birch bark, coriander, juniper, ginger, wintergreen, hops, burdock root, dandelion root, spikenard, pipsissewa, guaiacum, yellow dock, honey, clover, cinnamon, prickly-ash bark, quillaia, and yucca! Props to you if you're able to make the drink from scratch. Luckily, most grocery stores carry root beer concentrate, whose primary ingredient is artificial sassafrass root bark flavoring. (Real sassafrass root contains an aromatic called safrole, which, in 1960, the FDA decided was carginogenic and subsequently banned.) Yeast is typically added to carbonate the drink (and, in days past, to ferment it to make it slightly alcoholic). But rather than wait several days for the yeast to produce sufficient CO_2, you can use dry ice to the same end to make your own geeky root brew.

Ingredients

1 gallon hot water

2 cups sugar

2 tbsp root beer extract

1 to 2 lb dry ice

Equipment

Bucket or pitcher

Trivet, to protect surfaces from the extreme cold

Long-handled wooden spoon

Ladle

1 Measure out the hot water. Add the sugar. As with the lemonade, hot water makes the sugar dissolve more easily, and your final product will be more thoroughly carbonated.

2 Mix in the root beer extract. Your gallon of water will turn black and the whole room will smell like root beer.

3 Use a long-handled wooden spoon to stir the root beer mixture. Metal implements will work as well but can give you frostbite if you handle them without gloves. Also, metal implements tend to make terrible noises when they contact dry ice. Stir to dissolve the sugar, and add the dry ice.

4 The thick cloud of water vapor that will form over the bucket is truly amazing. You'll need to stir the mixture once in a while to keep it from freezing. The majority of the fog will dissipate, and the bubbles will begin to subside after about 15 minutes. At this point, the environment in the deep bucket is optimal for observing the "smoke ring" phenomenon (as described in the lemonade recipe), so watch closely. When the dry ice is gone, the root beer should be done—lightly carbonated and quite cold, almost to the point of turning to slush. You don't actually want slush, because it does not support carbonation. If your finished product is not cold enough yet, add some more dry ice and wait a few minutes. Serve with a ladle, and optionally garnish with a chip of dry ice. Serves 12.

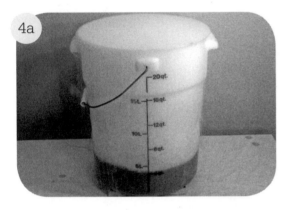

4a

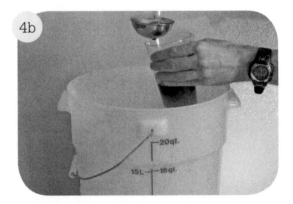

4b

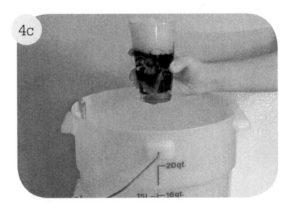

4c

[LEVITATING **BUBBLES**]

Here's a neat-o party trick you can do with dry ice. You'll need **dry ice**, a bottle of children's **glycerin bubbles**, and a **large bucket** or cooler.

Using a hammer and a dish towel to keep splinters from flying, break a few chunks of dry ice into the bucket. Wait a few minutes for a dense layer of carbon dioxide gas to form in the bottom of the bucket.

Next, blow some bubbles. Point them, as best you can, into the bucket. A few will, hopefully, go in.

We're accustomed to seeing floating bubbles as ephemeral entities, wandering unpredictably until they pop. But when bubbles enter the bucket, they change their behavior dramatically. They settle down and simply sit there. Ours floated from 6 to 12 inches above the bottom of the bucket. Some of them lasted longer than a minute. Occasionally, bubbles floating nearby would collide and form floating aggregates.

Why does this happen? When the frozen carbon dioxide sublimes (transforms directly from a solid to a gas), the cold carbon dioxide gas forms a layer in the bucket below the normal air above it. This layer consists of air that has been enriched with cold, dense carbon dioxide gas. Carbon dioxide gas is about one and a half times as dense as air under equal conditions.

The bubbles are filled with regular air, so they really do float on top of the carbon dioxide layer, much like a beach ball floating on water. While glycerin bubbles are normally short-lived, they can last for a surprisingly long time in this situation, since no solids, only gasses, are actually touching the bubbles.

*
See the Appendix for sources of dry ice.

Toasty Paws

STITCH MICROWAVABLE MITTEN WARMERS

*Most kinds of hand warmers commonly available on the market are disposable, made for one-time use. Our buddy **Rick Unger** showed us how to make reusable, cheap, and **easy-to-make mitts**. They're indispensable for anyone trekking into the Arctic, commuting through Brooklyn on a bicycle in January, or warming the bench for Mom during a soccer game.*

The key ingredient in these little hotties is rice. Inside a microwave, an electromagnetic field bounces the polar H_2O particles in the rice back and forth and knocks them into other molecules, thereby jostling them enough to heat them up. Once hot, the grains can take up to an hour to calm and cool down again, and your paws will stay extra-toasty in the meantime.

[PEEPING **TOMS**]

While you're hanging out near the microwave, enjoy a sophisticated game of marshmallow-Peep jousting. Stick a toothpick in the beaks of your two Peeps, and place them in the microwave facing each other, toothpicks *en garde*. Heat for one minute. The Peep that impales the other first wins.

Sew a 3- by 2-inch bag out of a **piece of fabric:** Begin with a piece of fabric a little wider than you want your finished bag to be and a little more than twice as long. (The extra width and length is your seam allowance.) You needn't fuss about neat cutting; just **snip** the edge and rip the fabric. Fold the piece in half with the inside out, and **sew** the two sides together. Then turn them right side out, and fill with uncooked **rice.** Tuck the remaining raw edges inside and stitch the opening shut.

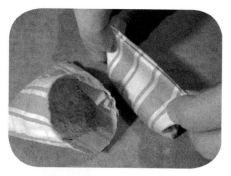

Microwave your finished mitten warmers between 30 seconds and 1 minute, depending on the strength of your oven, until piping hot. The rice will likely have a bit of moisture in it, so for the first few heatings, expect the warmers to come out of the microwave a bit damp. They will smell delicious! Pop them into your mitts right out of the microwave and enjoy toasty fingers.

If you don't have a sewing machine, use an **old but clean pair of socks.** Lop off the toes to make pockets with enough room to hold the rice. Fill the pockets with rice, and tuck in the raw edges, placing enough **fusible webbing** between to span the entire opening. **Iron** this seam together. Since microwaving these bags will melt the fusible webbing, you need to hand stitch along the seam. Socks tend to be quite stretchy and the fusible webbing helps stabilize the fabric while you sew it, though you could do without it if you prefer.

If you suffer from aichmophobia (fear of pointed objects) and don't want to sew anything at all, make a satchel out of an **old dish towel** or **T-shirt**. One way to make a satchel is to take a square of fabric and push some of it down into one hand and fill it with rice. Gather up the loose ends over the top and tie off with some **string**. Try not to make the satchel too tight; keep it a little slack. Then just trim off the excess and use.

If you never quite managed to learn the whole tying-string-together thing, use **rubber bands** instead. Wrap the rubber band as tightly as possible around the open end of the sock or satchel.

And if you really cannot be bothered to lift **more than a finger,** just put some uncooked rice in a bowl, heat it up in the microwave, and pour it into your mittens. The rice will be loose in your mitts and you'll probably loose some if you raise your hand to wave at a friend, but this is absolutely the fastest and easiest mitten warmer for keeping your fingers from freezing when you're outside on a winter's day.

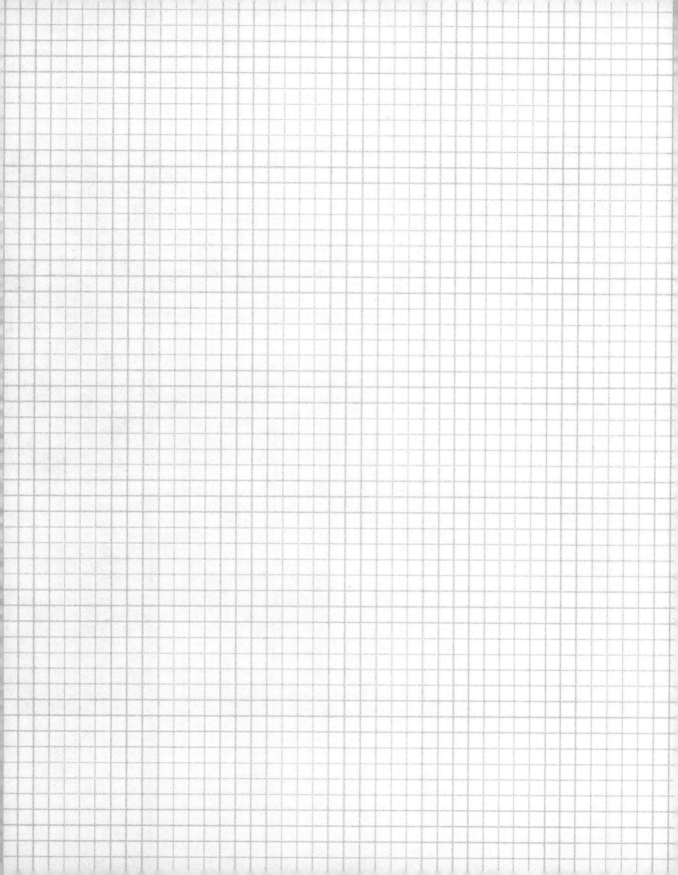

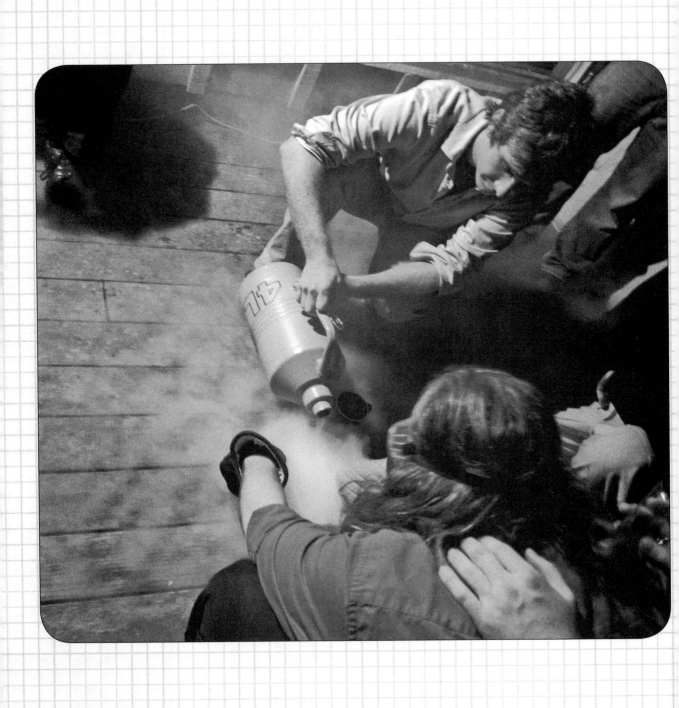

chapter 7

I Scream for Cryogenic Ice Cream

* *Freeze a far-out treat*

Making **ice cream with liquid nitrogen** is a fairly well-known gimmick, but the fact that LN2 actually makes a significantly better product made the DIY daredevil collective **Turkey Tek** think it was imperative we share this wealth far and wide.

A critical factor in ice-cream superiority is ice crystal size. Big crystals are not delicious. The top layer of ice cream that has thawed and refrozen in a carton in your freezer is gummy and tastes bad because when it refroze, it did so very slowly and with no stirring, which resulted in large crystals. This is also why you must constantly stir ice cream in a traditional ice-cream freezer. Decreasing crystal size improves not only ice cream's texture but its richness and flavor as well. Liquid nitrogen freezes cream so fast that the crystals have almost no time to form. As a result, the average crystal size in LN2 ice cream is far smaller than in conventional ice creams. This means that, if you choose your ingredients carefully, LN2 ice cream is very close to the theoretical supremum of all things tasty.

Cryogenic cream also facilitates the addition of liquor for flavoring that would otherwise inhibit freezing in traditionally made ice cream. The freezing point of ethanol at normal pressures is around −175°F. When you freeze ice cream using ice and rock salt, or even a state-of-the-art ice-cream maker, temperatures can't go nearly that low, and any attempt at hard-liquor ice cream results in ice-cream soup, like a well-chilled White Russian. The boiling point of LN2, however, is around −319°F, which is more than sufficient to do the job.

Below is a fancy recipe that involves cooking cream and egg yolks and guarantees The Real Thing. If you *really* want to cut corners, you can simply dump a bunch of milk, cream, sugar, flavoring, and liquid nitrogen into a bowl, and *voilà*. As always, experiment with flavorings. We also recommend making ice cream with hotter-'n-hell green chile powder. It's eyeball-burstingly good.

CAUTION: *Liquid nitrogen is quite cold and can easily freeze-burn if it comes in contact with skin. Be sure to wear insulated gloves and to not spill any on your shoes. Avoid wearing metal jewelry or watches that can quickly conduct cold to your skin. You should also wear eye protection, as LN2 boils rather violently at room temperature. Mix the ice cream in a well-ventilated area and store it in a ventilated container. Finally, wait until the ice cream has stopped smoking before you eat it!*

[GETTING THE **GOODS**]

The only hard part of making LN2 ice cream is obtaining the LN2. We found our tank when it fell off a lab truck; but if you're not so lucky, you can purchase it from a welding or medical-supply outlet, or get a buddy from a university lab to slip you some. Dewars (stainless-steel vacuum bottles) or Styrofoam coolers are safe transportation and storage vehicles for it. If you have a professional dewar, people are willing to assume you know what you're doing, and will sell it to you. If you show up with a Styrofoam cooler, you will be promptly turned away.

Scotch Whisky LN2 Ice Cream

Equipment

Heavy-bottomed pot

Small bowl

Large metal bowl

Wooden spoon

Ingredients

3½ oz of any type of peaty single-malt Scotch whisky

2 cups whole milk

2 cups heavy cream

4 egg yolks

1 cup sugar

1 liter liquid nitrogen

1 Place 2¾ oz of the whisky in a small heavy-bottomed pot. Standing back, ignite it with a long safety match to burn off the alcohol. When the flames die down, add the milk and cream. Place over low heat and stir occasionally until the mixture comes to a low boil, being careful not to let it scorch. Remove from the heat and set aside.

2 In a small bowl, whisk the egg yolks and sugar with an electric mixer until fluffy. Pour in half of the cream mixture and mix. Place the remaining half of the cream mixture above medium heat and bring to a boil again. Add the egg mixture and whisk for a couple of minutes. Remove from the heat and transfer to a large metal bowl. Add the rest of the whisky and thoroughly mix.

3 Slowly pour in the liquid nitrogen and stir constantly with a wooden spoon until the ice cream is too thick to stir.

*
See the Appendix or a source for ordering safety supplies.

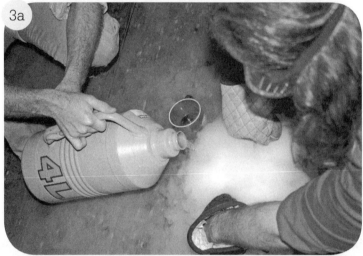

chapter 8

Warm Bud

Take a beer-can stove on the road

What's the one thing you should take to **a desert island**? If

you ask us here at **Hungry Scientist Headquarters**, make

it **a six-pack of Budweiser.** Why? First, to take the edge off.

Second, because you can use the cans to make a stove and fire

starter and then fry every fish in the sea.

Campers originally developed the beer-can stove because it's portable and featherlight. We know people who hiked the entire Appalachian Trail using them. Even at home, it makes a good backup in case of an emergency, and it is a pretty cool conversation piece. We had to try making this DIY Hall-of-Famer ourselves, and we discovered some helpful tips we hadn't come across anywhere else.

Equipment

A permanent marker

3 aluminum cans that do not have dents on the bottom; two empty, one full and refrigerated

A heavy hardcover book

An X-Acto knife

A quarter

A pushpin or nail

Foil tape

Metal polish (optional)

Heet lighter fluid or denatured alcohol fuel (found in most camping stores)

1 With a permanent marker, make 32 dots around the base of an empty can and 6 in the middle of the can. Puncture the holes with a pushpin or nail. This will become the "burner."

2 Use the book and the X-Acto knife to score a level line around the circumference of the can: Lay the book flat on a table and open it 7/8 of an inch up from the table. Insert the X-Acto knife base inside the book, with the sharp end of the blade sticking out of the top and facing the spine. Firmly hold the base of the knife with your thumb and the spine of the book with your index finger. Turn the can while holding the blade firmly to make the scored line.

3 Roughly cut through the can above the scored line to separate top and bottom. Make a diagonal cut up into the score line so that you can peel back the can along that line, leaving a level, smooth final top edge. (If you can figure out an easier way to make a perfectly level cut around the circumference of the can, let us know.)

4 Score the second can in the same way, this time 1⅜ inches from the bottom. Cut, peel, and separate the two parts of the can. This will become the "fuel reservoir."

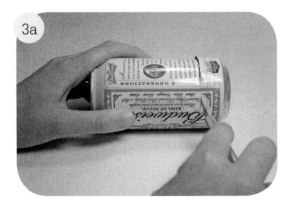

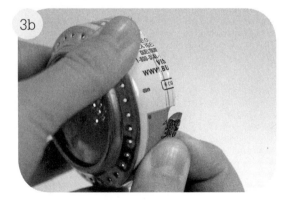

5 Score a strip from one of the top halves of the cans, 1½ by 7 inches, and cut and peel. The edges must be perfectly smooth.

6 Make a center fuel chamber by fitting the strip in a circle into the concave indentation in the base of the burner can (it will gently snap in). Tape the strip into a cylinder using foil tape.

7 Remove the center fuel chamber and cut three notches equidistant from one another into its top edge.

8 Using the full, cold can of beer as a mold, push the fuel reservoir over the end of the full can to stretch out the walls; this will allow the burner can to fit inside the fuel reservoir in the next step. Be sure not to bend or dent the edges of this reservoir.

9 Snugly fit the burner upside down into the fuel reservoir and press closed.

10 If you also want your stove to be a solar igniter for starting its own flame, simply polish the bottom of your beer can to a mirrorlike sheen. When the surface is nice and shiny, the can's concave shape can reflect the rays of the sun into a small pinpoint that has enough energy to light paper, dry twigs, or leaves on fire. The easiest way to get a mirror sheen on your can bottom is to use metal polish and a buffing wheel attached to a drill. However, if you are patient and willing, you can use crude chocolate with bits of cacao beans in it or toothpaste as metal polish. The finely ground beans in chocolate remove small amounts of the surface metal as they rub against the surface, bringing out the metal's shine.

9a

9b

9c

11 Test this outside on concrete or dirt! When you're ready to light the stove, pour fuel into the center punched holes, filling the stove about half-full. Then place a quarter over the center holes and add a little more fuel in the well. The quarter will block the fuel from flowing into the fuel chamber. Carefully light the pooled fuel. As it burns, it will warm and prime the stove. The burners should sputter to life after a minute of the priming puddle burning. You may have to repeat the priming step if it is cold outside, or if you did not add enough priming fuel to sufficiently warm the stove.

*
See the Appendix for information on more aluminum-can projects.

[LIGHT MY **FIRE**]

You can use the bottom of a beer can and a piece of paper to start a fire. First use something mildly abrasive to polish the bottom of the can. Crude chocolate works, but metal polish works much better. Position the shiny bottom of the can under bright sun. Its concave surface will bounce the light into a pinpoint-size spot an inch or so above the can; move the piece of paper above the can until you see a small bright spot of light (it helps to wear sunglasses when doing this). Focused long enough on flammable material, such as a piece of paper, and given bright enough sunlight, this solar concentrator will make a small fire come alive.

chapter 9

Tupperware **Party**

* *Put together a portable iPod boom box*

How much are **iPod docks** running these days? **Lee Von**

Kraus made us one for **$2.99**. Extra fresh.

N ot only will any Tupperware container from your kitchen suffice for this project, but almost all kinds of cut-able boxes will do. Speakers from an old computer that have an audio-in jack rather than an external power supply will allow the boom box to be conveniently portable. If you're using speakers with a wall adapter for power and you want to be able to walk around the block with your stereo on your shoulder, just wire up a battery pack that matches the voltage of the speakers' wall adapter. With some additional batteries, you could try adding CCFT lamps (made for computer interiors to make them glow) for some snazzy lighting effects.

Equipment

1 colored Tupperware container large enough to fit your two speakers	Mounting tape, such as Scotch brand
An iPod	A Dremel drill
Speakers	Packing Styrofoam

1 Take the lid off the plastic container and position the speakers and iPod inside it. Place the lid on top and trace where holes for the speakers should go. It helps to hold the speakers in place if you allow them to stick out of the front a little bit. Use mounting tape to hold the speakers in place as well.

2 Using the Dremel, cut the holes in the lid (and cut the Styrofoam packing pieces, if necessary) accordingly.

3 Drill small holes in the bottom corners of the lid and box to make a zip-tie hinge so that the lid doesn't pop off and let the iPod slide out the bottom. Drill handle holes at the top through which you can easily insert and remove your iPod.

4 Cut a circular hole in the lid to allow iPod control while it's in the stereo. Insert the speakers. Add the Styrofoam pieces and secure them with mounting tape.

5 Slide in your iPod, snap on the Tupperware lid, turn up the volume, and press PLAY.

chapter 10

Intergalactose
Scream

* *Make a milk-bottle megaphone*

What can you do with a milk bottle after you've **done your body good**? Minister of sound **Michael Zbyszyñski** uses it to call his **guests to the table**.

An electret microphone is the high-quality but inexpensive mic used in a telephone, and it can easily be made into a megaphone. It converts sound energy to electrical energy with a little magnet that moves when sound waves hit it. This electromagnetic disturbance translates into an electrical current. The LM386 operational amplifier jacks up the electrical signal enough to move the speaker. In turn, the speaker uses magnets to convert the amplified electrical signal into sound waves. The circuit in this project can be used to build a megaphone or amplifier for any kind of enclosure you desire. Recycling bin, here we come!

Equipment

Plastic milk bottle, thoroughly cleaned and dry

Electret microphone

LM386 audio amplifier

10k-ohm potentiometer

Knob for potentiometer

1k-ohm resistor

10-ohm resistor

220uF electrolytic capacitor

10uF electrolytic capacitor

0.1uF ceramic capacitor

3-inch 8-ohm speaker

Small PC (printed circuit) board

Small zip tie

Hookup wire

Snap connector for 9-V battery

9-V battery

¼-inch foam-core board, such as Fome-Cor

Soldering tools (soldering iron, solder, desoldering braid or pump, third hand or vice, lead trimmers, small pliers)

Multimeter

Hot-glue gun and hot glue

Heat-shrink tubing and heat gun

High-speed rotary tool with cutoff wheel, such as a Dremel

Drill with ¼-inch and ⅜-inch Forstner bits

Sharp knife

Metal ruler

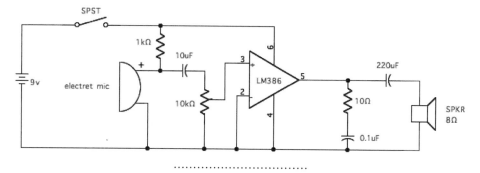

Circuit diagram for milk-bottle megaphone.

[CIRCUIT **SECRETS**]

Reading a circuit diagram is a nifty skill we think everyone should have. These diagrams are symbolic representations. They represent connections between components, not necessarily exact physical layouts. The symbols for the components in the megaphone circuit are as follows:

= resistor

= battery

= capacitor

= op amp

= potentiometer

= microphone

= switch

= speaker

Wires are indicated by solid black lines. Wires that connect do so at a solid dot, whereas wires that simply cross have no dot.

These wires are crossing, but not connected.

These wires are connected.

1   Using the circuit diagram on the previous page, solder the core of the circuit on a small (~ 1¾-inch square) PC board. Do not yet attach the battery, switch, potentiometer, microphone, or speaker. Instead, attach long (about a foot) lengths of hookup wire for these components, which will be attached after the circuit is mounted in the bottle.

2   The milk bottle should already be empty, clean, and dry. Use the cutoff wheel on the rotary tool to remove the bottom of the bottle. The same tool can be used to smooth any rough edges. Forstner bits make drilling plastic a little easier, but any ¼-inch drill bit will work. Drill holes for the potentiator and the switch. Put them where you like.

3   The bottle cap holds the microphone, and it unscrews so the battery can be changed. Lay out your electrect microphone and battery snap connector so that they both fit inside the cap and will not interfere with the cap's threads when a 9-V battery is attached. The next step is to drill three holes in the cap: a large one for the microphone and two smaller ones for the zip tie that holds the battery snap connector. (See photo for placement. The microphone is the silver blob on the upper-right side in the photo.) Find a drill bit that is the same size as the microphone and drill the hole. Use a small bit to drill holes for the zip tie that will help hold the snap connector and battery in place.

4   There will be more than enough room in the bottle for the circuit board. Run a generous line of hot glue along the bottom edges of the board. Feed the wires for the microphone and battery/switch through the bottom of the bottle and out the top. Then put the circuit board into the bottle and press it against the side of the bottle to secure it with the glue. Be careful not to cover the holes for the switch or knob. When you are finished, the board should be glued in the middle of the bottle, with battery and microphone wires coming out the top, speaker and potentiometer wires coming out the bottom.

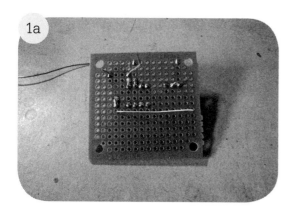

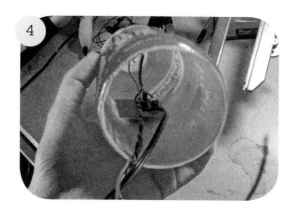

5 Find the three wires for the potentiometer (pot) feeding out the bottom of the bottle from your circuit board. This pot will function as a volume knob. Attach the wires to the pot using solder and heat-shrink tubing. Remove the nut and washer from the shaft of the pot, and feed the pot through the hole from step 2. Thread the washer and nut back on the shaft and tighten to secure. Loosen the setscrew on the knob (if there is one) and attach the knob to the shaft. Tighten.

6 The switch goes between the positive terminal of the battery snap connector and the rest of the circuit. Find the appropriate wires, coming out of the top of the bottle, and use solder and heat shrink to attach the switch. As in the previous step, remove the nut and washer, thread through the hole, and tighten in place.

7 Dry fit the microphone into the hole in the cap; it should be pretty snug and the element should be facing out. Use hot glue on the back to fix it in place, being careful not to bury the leads. Thread a zip tie through the holes in the cap. Cover the back of the battery snap connector in hot glue, stick in place, and secure with the zip tie. Cut off the extra length from the zip tie.

8 Solder and heat shrink the positive wire from the battery connector to the wire for the switch. The ground from the microphone and the battery are both soldered to the lead for the circuit's ground. The output of the microphone attaches to the last wire.

9 Solder and heat shrink the speaker to the remaining two wires coming from the bottom of the bottle. The speaker should rest about an inch from the bottom of the bottle. A seam in the label of this bottle is approximately the right place. Cut 4 (1-inch) squares of foam-core board. Cover the backs of the squares with hot glue, and stick them, equidistant from one another, to the inside of the bottle so that their bottom edges run along the marked seam. They should all be the same distance from the

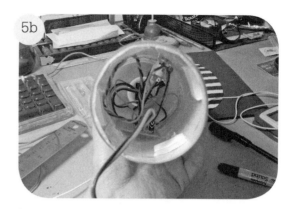

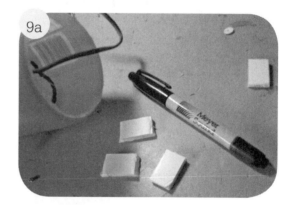

bottom of the bottle. Cover the edges nearest to the bottom of the bottle with hot glue, and slide the speaker into the bottle. Be careful that none of the wires gets stuck to the glue. Gently press the speaker against the foam core for a few seconds until the glue sets.

10 After all the glue dries, snap a 9-V battery into the cap. Carefully screw the cap on; there should be enough extra wire so that there is not too much strain on any of the connections when the cap turns. Press the button on the side and talk into the microphone.

11 The megaphone works best with the volume knob approximately centered. If no sound comes out, try adjusting this knob. There should be a small pop when the switch is pressed. If not, the battery may be flat or some connection may be broken. Correct it, and start yelling.

chapter 11

Hot **Box**

* *Build an outdoor roasting contraption*

Several years ago, a traditional Cuban outdoor roasting box called La Caja China appeared on the U.S. market. Its **large size** allows foods that don't fit in a conventional oven to cook quickly inside it. And we all know a whole roasted bird or piggy is a **carnivore's delight**. Our old friend **Ryan Horan** got grizzly, fashioned one of these boxes from scratch, and spent the evening licking his chops.

A roasting box has a simple design. A charcoal fire built on top of a metal-lined plywood box provides the heat source. Waves of infrared energy reflect off the shiny steel, generating heat that moves equally well in all directions rather than solely up. This is how the roasting box provides a controlled, evenly distributed, gentle heat that cooks the food.

We've modified the design of the commercial roasting box to make it scalable according to the size of what's cooked inside it. The one demonstrated here is roughly 2 × 1½ × 3 feet, which is perfect for cooking two whole chickens and lots of roasted vegetables and which can be made from a single sheet of standard-size plywood. It's important that all the foods going into the box have similar cooking times or are able to withstand extended cooking. Potatoes and other root vegetables wrapped in foil work well; more delicate vegetables can't hold up in the heat.

Equipment

1 (4×8 foot) sheet of ½-inch-thick plywood

Small nails or screws

2 (3½×2½ foot) and 2 (2×2 foot) pieces of new, clean, shiny sheet steel

Heavy-gauge foil

Staple gun or metal tacks

4 concrete blocks or bricks

2 (20-lb) bags of charcoal

Digital oven meat thermometer

2 barbecue forks

Heavy aluminum garbage can

Sand

Shovel

Heavy-duty work gloves

1 Cut the plywood (or have it cut in the store when you buy it) into one $3\frac{1}{2} \times 4$ foot, two $2 \times 1\frac{1}{2}$ foot, and two $3 \times 1\frac{1}{2}$ foot sheets. Atop a concrete, dirt, or sand surface outside, nail or screw together four pieces of plywood to create a four-sided box with no top or bottom. Make sure the corners fit tightly so that heat doesn't escape. Line every surface, inside and out, with clean heavy-gauge aluminum foil, shiny side facing in. Because the metal serves to reflect and radiate the heat, it is important that it's shiny and covering every surface. Use a staple gun or metal tacks—but not glue—if the foil doesn't stay in place.

2 Construct the bottom of the oven by placing one strip of $3\frac{1}{2} \times 2\frac{1}{2}$ foot sheet steel over the $3\frac{1}{2} \times 4$ foot sheet of plywood. Do not attach this bottom to the box. Place the entire bottom on top of four concrete blocks or bricks. (The air gap this creates will insulate the box and keep it hotter. We learned this the hard way, spending hours with our oven at subcooking temperatures before we propped it off the ground.)

3 Place the other $3\frac{1}{2} \times 2\frac{1}{2}$ foot piece of sheet steel on top of the box to make the lid. Bend it across two opposing edges so that it doesn't slide off.

4 Construct two fire pans out of sheet steel. Take the two 2×2 foot sheets and fold up their sides to create a lip that will contain several pounds of charcoal.

5 Prepare your food. Since you're not using wood and the food won't be flavored by wood smoke, make sure it's thoroughly marinated, brined, or seasoned. Before you place the meat in the box, insert a digital thermometer into its middle without letting it touch the bones and run the thermometer's cord out the top and over the side of the box. (Once the lid is closed, **you must not peek** into the box while the food is cooking or the distribution of the heat will be thrown off; so the thermometer is the only surefire way to tell when the meat is done.) Put the food in the box; cover with the lid and then the fire pans.

6 Evenly spread the charcoal between both fire pans. Light the coals. Then wait . . . and wait . . . and wait. . . . While the meat cooks, clear the ash as it builds up on the box so that it doesn't insulate the coals and keep heat from radiating. You can use a hair dryer to blast off the ash and stoke the coals, or just blow on them if you have strong lungs. Our chickens were

ready when their internal temperature reached approximately 165°F. When you've deemed the meat done, remove it with barbecue forks, along with any veggies. Let it cool for at least fifteen minutes, and while it's cooling, use the spade or shovel to remove the hot coals to a garbage can full of sand. When you're done, carve up the meat, serve, and enjoy!

*
See the Appendix for more resources on cooking with a Caja China.

Salt and Pepper Scooter

Every gathering of Hungry Scientists rightfully takes place around an enormous dinner table. The table needs to be big enough to hold a massive spread of food and still have plenty of space for tinkering, both during and in between meals. In lieu of a Lazy Susan to pass items around the table, this is a handy little vehicle to scoot salt and pepper shakers around glasses and gadgets to your friends at the opposite end.

Take apart a **long windup toy car**. Fit a pair of **cylindrical salt and pepper shakers** on the car's chassis. Cut two (1-inch) wide strips of aluminum from a **soda can** that are long enough to wrap around the shakers. Connect the end pieces of each together with **foil tape** or **epoxy**. **Glue** the aluminum housings to the car chassis with epoxy. Insert the shakers, and send them flying!

mini project.

chapter 12

Edible Origami

Fold and fry crane-shaped croutons

According to our research, never in the history of **the art of paper folding** (it was first documented in Japan in the early 1600s) has anyone made edible origami. Naturally, the delicately dexterous **Lenore Edman** figured out how to do it.

Wonton wrappers are strong and flexible, making them ideal for folding. We recommend that you practice making a crane with paper before you try with wonton wrappers. If you've made cranes before, you'll do just fine. Perched atop an Asian salad, a deep-fried wonton crane is truly a pièce de résistance.

Ingredients

1 package of thin, square wonton wrappers

Vegetable oil for frying

Deep-fry or candy thermometer, or a chopstick

[FOOLPROOF **FOLDING**]

➤ If you fold paper back and forth enough times, it will break; the same is true for wonton wrappers, only more so. You can do a little bit of re-gluing with water, but not too much.

➤ Absolutely square is ideal for origami paper, but since wonton wrappers are a little bit stretchy, you can usually make them do your bidding even if they aren't perfectly shaped. They can also be modified with a pizza cutter.

➤ While you are folding, keep the rest of the wrappers sealed in a plastic bag or covered with a damp cloth. If they dry out they become brittle.

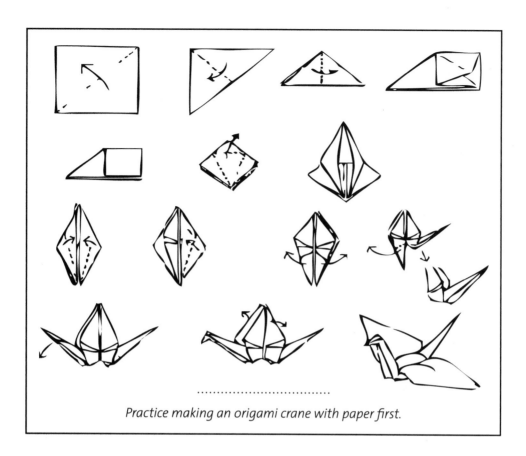

Practice making an origami crane with paper first.

1 Fold a square in half to form a rectangle.

2 With the folded edge at the top, fold over the upper-left-hand corner to meet the bottom edge, marking a diagonal fold from the center to the lower-left-hand corner. You should end up with a trapezoid.

3 Fold (push) the pointy lower-left-hand corner *inside* the square portion on the right (between the two layers of wonton) to meet the lower-right-hand corner. You should now have a square ¼ the size of the original wonton. Fold the top (there are three) lower-left-hand corner back to meet the upper-right-hand corner, so that the entire square is only four layers thick.

4 Here's the tricky bit: If at this point you pre-fold the edges to form a kite shape, you may crease the dough to the breaking point. Instead, it is best to lift up one layer of dough and gently guide it into the rhombus shape with your fingers.

5 Turn over and repeat.

6 Fold the sides in to taper the head and tail. Turn over and repeat.

7 Open one side slightly and fold up to form the head or tail. Press the dough firmly together at the top; there are a lot of layers there and they need to be coaxed into shape.

8 Fold up the other side and fold down the head. What with all those layers, the head isn't going to fold neatly unless you were a perfectionist in all of the previous steps. Treat it like clay; give it a little bit of water and firm pressure. If you plan to deep-fry them, fold down one wing of each crane. Keeping them relatively flat makes it easier to gently flip them with tongs as they are frying.

9 If you choose to microwave them, open the wings and press down on the back gently to expand it. Microwave until the surface bubbles. Microwaved cranes keep their three-dimensional shape beautifully. They stay a pale white that contrasts well with salad greens. They are rather bland, however, and taste much like a dried-out flour tortilla.

10 To deep-fry them, pour an inch of vegetable oil into a high-sided, heavy-bottomed pot and place it over medium-high heat. When the oil reaches 375°F on a deep-fry or candy thermometer, or when bubbles cover a chopstick when it's stuck into the hot oil, gently grab one crane at a time with tongs and lower it slowly into the oil. Deep-fry it briefly on each side until it turns golden brown on the edges. (It is fine for the center to be a little pale.) Remove from the oil and let cool on a paper-towel-lined plate.

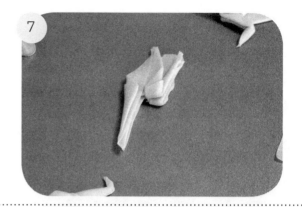

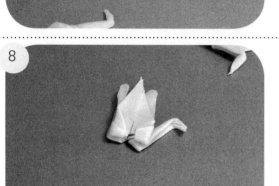

*
Try other traditional origami shapes, such as the throwing star, iris, balloon, or frog.
Some shapes may even be able to hold wonton fillings, such as the balloon.

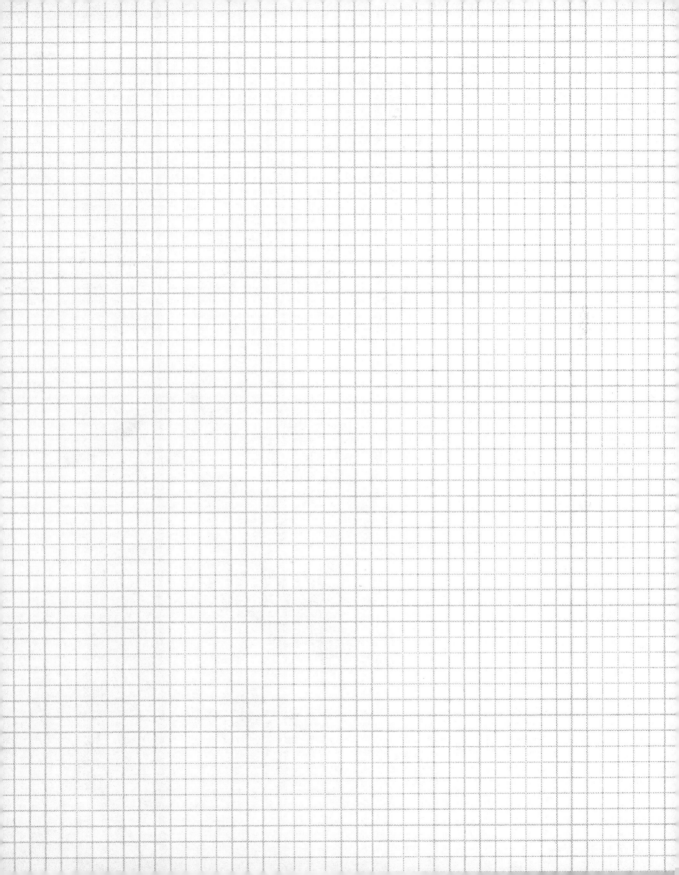

chapter 13

Gravy Train

*** Construct a colossal meat-juice fountain*

Any holiday feast is an excuse for Hungry Scientists to **attack the kitchen** and realize their **wild feast dreams**. Case in point, a fountain that can accommodate 10 gallons of gravy—chez **Turkey Tek Collective**—is the paragon of Thanksgiving opulence. Once you experience its gravyliciousness, you'll never go back to that tiny gravy boat.

A wide variety of impeller pumps are available on the market, but they pulverize giblets and jam when thick gravy passes through them. For more reliable functionality, we based our design on the peristaltic pump, a medical device used to move blood during surgery without damaging blood cells. Two successive rotor rollers gently pump pockets of gravy through a soft rubber hose in a process akin to squeezing a never-ending tube of toothpaste. Powered by an electric drill, the pump comes to life and provides a gushing torrent of delicious, piping hot gravy. Bliss.

Equipment

Handsaw

Wood screws

Electric drill

1 (8 foot) 2×4

2 ball bearings

2 (16-inch-diameter) circular pieces of particleboard or plywood

6 plastic rollers

1 (foot-long) threaded rod

24 bolts and washers (to match threaded rod)

A strip of sheet metal as wide as the rollers are tall

7 angle brackets

Construction adhesive

3 feet of 2-inch-diameter flexible rubber hose

Zip ties

Hose clamps (optional)

Dimmer switch

Extra-large stockpot

Caulk

Sink drain fitting

Hot plate

Chrome pipe

1 (4×8 foot) sheet of ½-inch plywood

2 large decorative bowls

1 First, build a plywood enclosure to hide the stockpot reservoir and hot plate: Size and geometry will vary with pot size. Using the handsaw to cut the plywood, cut two identical trapezoids with flat bottoms and a sloped side on the top. Cut four cross pieces out of the 2×4 long enough to leave plenty of clearance around the stockpot. Cut two more plywood panels to form the front and sloped top of the enclosure. Assemble all of the panels tightly with wood screws. You might be able to get away with using an old unwanted file cabinet as an enclosure.

2 Cut out two 16-inch-diameter circles from either plywood or particleboard. Drill a hole through the middle of each circular disc and six equidistant holes around their perimeters. Assemble the rollers, bolting them between the two circular discs, and bolt the threaded rod through the middle hole with enough length on each side to extend through the 2×4s that you will be constructing the frame from in the next step.

3 Cut a 30-inch square of plywood and one 30-inch and two 9-inch lengths of 2×4 to form the frame of the peristaltic pump. Drill holes in the middle of the 30-inch 2×4 and the middle of the 30-inch-square piece of plywood large enough to accommodate the bearings. Temporarily assemble the frame and roller assembly to make sure it all fits together properly. (Refer to photo 6a to see how this will look when fully assembled.) Tighten the drill chuck onto the threaded rod and give it a whirl. If your bearings are aligned, your roller assembly should rotate relatively easily.

4 Drill holes along the edge of the sheet metal backing plate for the six angle brackets. Glue the rubber hose tubing to the metal backing plate with construction adhesive. Take your assembled frame apart and bolt the rubber hose metal backing plate assembly onto the 30-inch-square plywood using the angle brackets, being sure to allow enough room to fit the 16-inch-diameter roller assembly. The rollers should come into firm contact with the metal plate in order to seal the rubber tube but still allow the whole thing to rotate freely. Mount the roller assembly, bearings, and reassemble your wooden frame from the pieces cut in step 3.

5 Secure the drill to the backside of the plywood enclosure using another angle bracket and zip ties as necessary, and tighten the drill chuck onto the axle. Use another zip tie to permanently hold down the trigger on the drill, and attach a lamp dimmer switch to it to provide precise control over the motor speed. In the end, it will probably be necessary to run the drill at a considerable clip. AC motors have relatively little torque at low speeds and may otherwise stall under the pressure of 10 gallons of delicious gravy.

6 Once the housing and pump are assembled, plumb the rest of the fountain. Drill a hole in the side of the stockpot large enough to insert the sink drain fitting. Seal it with caulk and arrange the pot on the hot plate. Create a spout by attaching the chrome pipe at the top of the fountain using a metal pipe hanger or more zip ties. Seal the pump hose to the stockpot and

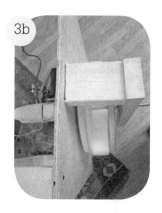

spout with zip ties or hose clamps. Mount one or more decorative bowls on the sloping plywood to provide an artful cascade of gravy from the spout. Drill a hole in the bottom bowl and sloping plywood so that gravy flows through, returning to the stockpot below. You are now ready to enjoy the delights of 10 gallons of gravy!

6c

6d

Gravy

Makes 10 gallons

Equipment	Ingredients
Heavy-bottomed roasting pan	15 to 20 lb chicken bones and backs
Large stockpot	Giblets (optional)
Large soup pot	5 lb yellow onions, peeled, and cut into 16ths
	3 lb unpeeled carrots, scrubbed and cut into 1-inch lengths
	1 head celery, washed and cut into large pieces
	Handful of fresh thyme
	Half bunch of parsley
	2 bottles white wine
	Large pinch of black pepper
	2 or 3 bay leaves
	2½ lb butter
	2½ lb flour

1 To make the stock, heat the oven to 450°F. In a heavy-bottomed roasting pan, toss the bones and vegetables with vegetable oil and roast (in batches, if necessary) for about 30 minutes, or until they reach a rich dark brown color. Remove them from the oven, transfer to a stockpot, add the

herbs, and cover with cold water. Deglaze the roasting pan with white wine. Add that to the stockpot and bring to a boil. Lower the heat and simmer for 1½ hours, periodically skimming the surface. Strain the stock, chill, and reserve for gravy.

2 To make the roux, or thickening paste, melt the butter over medium-low heat in a large soup pot. Add the flour, stirring constantly, until smoothly combined. Cook, watching like a hawk, until the mixture is aromatic and brown. Add a cup of cold stock to the hot roux to make a thick sauce.

3 Stir until smooth, then add the rest of the stock to the pot. The roux will not thicken the liquid until the mixture comes to a boil. Once the pot boils, lower the heat to a simmer. Add the giblets if desired. Simmer for 10 to 15 minutes and season to taste. Seasonings can still be added after the gravy is loaded into the fountain.

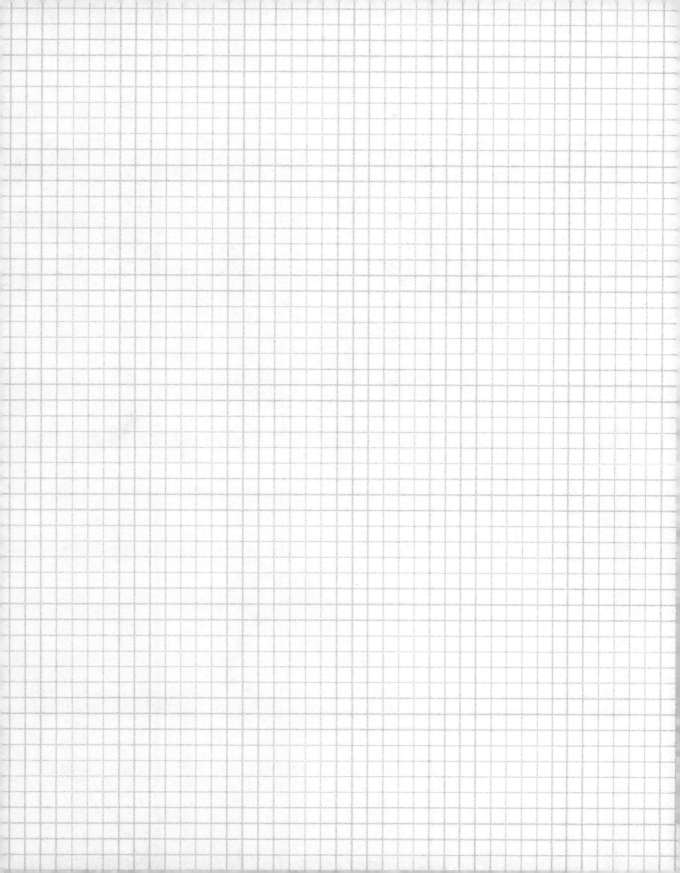

chapter 14

Mallow Ammo

Whip up launchable marshmallows

We dream of **raining marshmallows**. And when we Hungry Scientists dream, we tend to put our **mad dreams to life**. We asked our gourmet guru **Brigit Binns** to be our guide, and this is what she led us to.

Marshmallows are basically a semipermanent form of foam. When a protein substance such as gelatin or egg white is whipped with sugar syrup, creating a frothy mixture, the protein molecules fill the spaces around the air bubbles and fix the foam's aerated structure. Homemade marshmallows (the old-school version is made with juice from the marshmallow root, but we've spared you that here) are worth making because they're fluffier and lighter than their store-bought sisters: all the better for launching.

Equipment

12 × 17 inch rimmed nonstick baking sheet or jelly-roll pan

Large heavy saucepan

Baking parchment

Handheld or stand electric mixer

Candy thermometer

Surgical gloves or tight-fitting rubber gloves

Vegetable oil

Large knife

Large cutting board

Round or shaped cookie cutters (optional)

Ingredients

Baking spray, such as Baker's Joy

½ cup powdered sugar

½ cup cornstarch

3½ envelopes (2 tbsp plus 2½ tsp) unflavored gelatin

½ cup cold water

2 cups granulated sugar

½ cup light corn syrup

½ cup hot water

¼ tsp salt

2 large egg whites

2 tsp flavoring oil, such as peppermint or cinnamon (optional)

3 to 5 drops pink or green food coloring (optional)

1 Read through the recipe and assemble all the equipment and ingredients needed: Once you start, you've got to be prepared to move quickly.

2 Spray with baking spray the base and sides of a 12×17 inch rimmed rectangular nonstick baking sheet or other rimmed plate, and then line it with a sheet of baking parchment. In a bowl, combine the powdered sugar and the cornstarch and mix evenly. This mixture will be used throughout the recipe. Spray the parchment with baking spray and dust it evenly with the powdered sugar mixture, using a sieve to ensure even dusting.

3 In a very large metal bowl, sprinkle the gelatin over the cold water and let it stand to soften.

4 In a large, heavy saucepan combine the granulated sugar, corn syrup, hot water, and salt. Place over low heat and cook, stirring with a wooden spoon, until the sugar has dissolved. Increase the heat to medium-high and bring the mixture to a boil; simmer gently *without stirring* until a candy thermometer registers 240°F; this should take about 9 minutes. (The temperature will hover around 200°F for a while as the water boils off. Keep an eye on the thermometer because the temperature will shoot up suddenly.) Pour the hot sugar mixture over the gelatin mixture, stirring until the gelatin is dissolved.

5 Immediately, with a handheld electric mixer (or in a stand mixer), beat the thin mixture (5a) on high speed until it's white, very thick and gluey, and nearly tripled in volume (5b), about 7 minutes.

6 Working quickly, wash the beaters thoroughly in hot water, and in another large bowl, beat the egg whites on high speed until they just hold stiff peaks. Add the beaten egg whites, flavoring oil, if using, and food coloring, if using, to the sugar mixture and fold just until evenly combined.

7 Scoop the mixture onto the prepared baking sheet; then quickly put on the rubber gloves and oil them well. With a light hand, pat the marshmallow into an even layer; then sift about ¼ cup of the powdered sugar mixture evenly over the top. Set aside some of the powdered sugar mixture for the next step. Chill the marshmallow, uncovered, until firm, at least 3 hours and up to 1 day.

8 Pick up one edge of the parchment paper and slide the whole sheet of marshmallow onto a large cutting board. With a large knife, trim the edges of the marshmallow and cut the marshmallow into roughly ¼-inch cubes. (Or, use a small shaped cookie cutter to make birds, trees, stars, or whatever else your heart may desire.) Sift the remaining powdered sugar mixture into a large shallow bowl and add the marshmallows in batches, tossing them to coat evenly. Marshmallows will keep in an airtight container at cool room temperature for 2 weeks. Yields about 120 (¼-inch) marshmallows.

Lego Catapult

Legos are part of a universal system: Every single one of the more than 20 billion pieces of Lego bricks manufactured every year is compatible with all of the rest. Accordingly, we're working on a complete kitchen made out of Legos. When we take breaks, we watch Lenore Edman play with her Lego launcher.

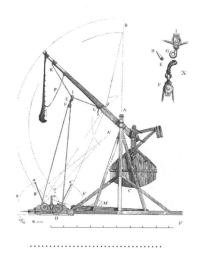

The model: medieval trebuchet.

Marshmallow-launching trebuchet.

Ammo and sling.

A marshmallow is released.

*Many marshmallows have
been released.*

chapter 15

Living **Loaf**

* *Catch wild yeast and bake a* boule

Bread bakers are **hard-core biochemists**: They must under-stand how to manipulate microorganisms in order to perform their magic. Master baker **Ryan Horan** taught us how to **harness wild yeast** and, without much effort, turn out deli-cious loaves of bread.

In our baking curriculum, Bread Biochemistry 101 requires catching your own wild yeast. Of the many kinds of microorganisms floating about us, yeast is one of the few that feeds on complex carbohydrates. Bacteria tend to eat simple sugars and proteins, which is why juice molds when left out of the refrigerator for a few days, whereas yeast in a flour-water mixture starts to ferment. With a simple mixture of flour, water, and a wee bit of sugar, a biochemical trap can be set up to tempt yeast's picky palate and discourage the bacteria that can't handle the complex food. The yeast cells begin to chow on the starches and sugars, expelling carbon dioxide gas as they go, dividing and multiplying as they become better and better fed. The more yeast cells there are and the gassier the lot, the higher the dough that they eventually infiltrate will rise and the more heavenly the resulting bread.

There are many different strains of yeast, and each has a different flavor. Wild-caught yeast makes bread taste peculiar to where you live. The flavor of a *boule* (European-style round bread) from Brooklyn is going to be as subtly nuanced compared to a sourdough from San Francisco as those cities' respective hipsters are. Sometimes wild-caught yeast bread will turn out mildly sour because the air in a given location is rife with acid-forming bacteria that sneak into the fermenting yeast starter. Acid and the protein-chomping bacterial enzymes will retard the yeast's gas production, thus making the dough less elastic and denser. Just give the dough plenty of time to rise. If you're patient, you'll hardly need to knead it.

[BREAD OVEN **LIFE**]

Ovens truly affect the measure of good bread. The sad truth is that after any adventurous home baker has taken the time to make his own bread, his conventional gas or electric oven does not have the ability to honor the endeavor and bake it as well as it should, and can, be baked. Bread benefits from high, evenly conducted heat. Modern ovens have thin walls that don't retain heat, generate even heat throughout, or get very hot in the first place (500°F, at most). Traditional dome-shaped bread ovens made out of clay, stone, or brick retain the almost 900°F heat generated in an initial blast from a wood fire built beneath the bottom stone. The blast radiates directly up into the loaf and enhances the crispy, browned crust and toasty flavor caused by the Maillard reaction (see page 2). You can buy expensive baking stones, ceramic inserts, or cast-iron pots to simulate a good conductive baking oven, or you can create a cheap and excellent facsimile of one by lining your oven with scrap materials such as quarry tiles and fire bricks.

Equipment

Medium-size metal or ceramic bowl

Cookie sheet or TV dinner tray

Large bowl

2 unglazed quarry tiles just a tad smaller than the rack in your kitchen's oven

Enough fire bricks (untreated, safe for baking) to line the back and side of your oven

Small cast-iron pan

Cheesecloth or medical gauze (optional but recommended)

Baking thermometer (optional but recommended)

Ingredients

4 cups unbleached white flour

1 cup distilled water

1 tsp sugar

1½ tsp salt

1 cup warm tap water

Olive oil and garlic, for garnish (optional)

1 Mix the distilled water and 1 cup flour and a teaspoon of sugar in a medium-size metal or ceramic bowl. Cover with cheesecloth or medical gauze and set aside in a warm environment (as close to 70°F as possible) and, ideally, where it will be touched by outdoor air—either in an open window or outdoors—and where it will be undisturbed. It should turn into a seething, bubbling mass of yeast cultures in about five days. This is called the starter, and it should give off a pleasantly sour, almost beery smell. If the starter smells fishy or putrid or especially if you notice the growth of mold on your culture, toss it and start over.

2 In a clean, large bowl, mix the remaining 3 cups of flour, salt, and the warm tap water. Mix in enough yeast starter to make the dough the consistency of very thick pancake batter—sticky and wet but not watery and runny.

3 Cover the bowl with a clean dish towel and place it in a cool spot where it won't be disturbed. The mass of dough will begin to puff up as it comes alive. As time passes, two main biochemical reactions occur in our mixture of water, flour, yeast, and salt. One, the yeast ferments the natural sugars and starches in the dough, resulting in a pleasant, full, and savory flavor. Two, and this is the important detail, the gluten protein in the flour begins to unravel and combine to give the necessary strength to the dough that is typically accomplished through kneading. Given enough time, your microorganisms will do the work of kneading for you.

1

2a

2b

2c

4 After at least 24 hours (we have let dough sit for as long as five days), turn the dough out onto a well-floured surface. Fold the mass onto itself. Rotate it and repeat the process several times, until the dough begins to develop real body. Doing this provides a bit more structure to the flour's network of protein molecules, allowing it to trap air in the dough. You should be able to smooth the dough into a dome by pulling the surface layer down and into the base of the mass, leaving a smooth surface. Cover the dough once more with a clean towel (dust it with flour, if needed, to keep it from sticking) and let it rest for an hour or two. The yeast has reached fermentation perfection once the dough has at least doubled in size, retaining a hole when you poke it with your finger.

5 At least half an hour before you're ready to bake the bread, prepare your oven. If you have quarry tiles and fire bricks, line your oven racks and walls. Place an empty small cast-iron pan to one side in the bread oven. Crank up the oven thermostat as high as it'll go.

6 When the oven is as hot as it's going to get, slip a floured cookie sheet beneath the dough and transfer it to the bread oven. Pour a cup of water into the cast-iron pan. The clouds of vapor produced will coat the dough with a thin film of water that will seal in the yeast's gas as it expands with the heat, allowing the loaf to puff up to an optimal height. When the film dries, it will create a beautiful, glossy, crackly crust.

7 After 30 minutes of baking, remove the cast-iron pan. Bake for another 15 to 25 minutes or until the interior of the loaf registers 200 to 205°F on a baking thermometer. Gently remove the bread with dishrag and allow it to cool for at least one hour.

chapter 16

Basement Bacchanalia

* *Ferment pomegranate wine*

Making moonshine of one sort or another is one of the most **ancient DIY endeavors**. These days, though, with all the modern equipment that most people are convinced they need to buy in order to make their own wine, homemade juice is hardly cheaper than store-bought bottles. We asked our oenophile friends **Jennifer Botto** and **Ryan Horan** to make us some fine vino with what they could find at home. They came back with **something truly mythic:** a bottle of pomegranate wine.

Pomegranates are filled with tannins—the compounds that in grape skins impart a dryness to the mouthful of red wine—and make a fine wine when their juice is fermented. They're also teeming with antioxidants, as advertisers of pomegranate juice have been so keen to let the public know, so this wine is theoretically *very* healthy to drink.

Making wine with pomegranates is not much different from making it with grapes. The same three elementary stages are crushing the fruit, fermenting the juice to convert its sugar into alcohol, and aging the wine to develop its flavor. The beauty of making wine at home is that every batch is different; the final product can range from sweet to sour or be more or less tannic. The flavor of this kind of wine tends to be quite potent, however, so while the wine is lovely straight up, it also mixes well in spritzers, margaritas, and granitas.

Equipment

Large enamel pot

3-gallon plastic bucket

Freshly laundered dish towel

Clean glass jugs, totaling 2 gallons in volume

Water-lock stoppers, old corks, or cotton stoppers

Plastic siphon tube or plastic funnel

Plastic or wooden spoon

Empty bottles of one sort or another

New corks (optional)

Corker (optional)

Ingredients

6 ripe pomegranates

3 lemons

2 gallons water

4 to 6 lb sugar

1 slice toasted wheat bread

2 (¼ oz) packages Fleischmann's Active Dry Yeast

½ cup warm water

Makes approximately 2 gallons

1 Prepare a simple syrup by placing the sugar into a large enamel pot. Depending on how sweet you would like your wine, use between 4 and 6 pounds of sugar to taste. Add 2 gallons of water. If you do not have a pot large enough to hold all the sugar and water, you may split them equally between different pots.

2 Boil the water until the sugar is fully dissolved and the mixture becomes slightly syrupy. (It will not reach the thickness of true simple syrup because the amount of water used in this recipe is proportionately greater than in a typical recipe.) Let the mixture cool until it becomes lukewarm.

3 Prepare the fruit base: Because the outer skin of the pomegranate will mingle with the seeds in the water, make sure to wash the outside of the fruit thoroughly before you start, even if the fruit is organic. Seed each pomegranate by cutting it into quarters, immersing it in water, and carefully separating the seeds from the white membrane under the water. The white membrane will float on top, and the seeds will sink to the bottom. Don't worry about the small amount of juice lost to the water as you seed; you can reserve some of this water to feed to your yeast starter a bit later.

4 Collect and strain the seeds and place them into the 3-gallon plastic bucket, crushing them with your hands as you go. *Do not let the fruit (or yeast or wine) ever come in contact with metal.*

5 Peel the lemons and discard half of the peels. Cut the lemons in half and squeeze them (but not their seeds) into the mixture. Toss in the reserved peels; they will impart a slight bitterness that will round out the finished wine. The mixture you've just created is called the must. Its sugar and acidity offer the perfect environment for yeast cells to propagate.

6 Prepare your yeast starter: Dissolve the 2 packages of Fleischmann's Active Dry Yeast in the ½ cup warm water. You may use the same water you used for seeding the pomegranates, as the yeast needs a small amount of sugar to feed on initially and will be energized by the residual juice left in the water from the seeding process.

7 After about 10 minutes, you should see foam on the surface of the water, indicating that the yeast is properly hydrated. Cut a piece of toasted wheat bread in half and spoon some of the yeast mixture on top of one half. Let the half piece of toast float in the starter. The carbohydrates in the bread will help to feed the yeast. After a few minutes, add a few tablespoons of the lukewarm simple syrup and a few pomegranate seeds to feed the yeast.

8 Start the primary fermentation: Make sure the temperatures of all your mixing ingredients at this point are lukewarm. Pour the simple syrup into the must and mix with a plastic utensil.

9 Check on the progress of the yeast starter. The yeast byproduct (carbon dioxide) should now be erupting violently. Place the half piece of toast from your yeast starter into the must, followed by a few tablespoons of the starter. Every few minutes, continue adding just a few more tablespoons of the starter around the toast. Watch and listen for signs of fermentation.

Alternatively, add a few tablespoons of must to the yeast starter. This will balance the environment to match the must. Aerate the must by mixing and flipping the toast around. You should notice bubbles forming under and around the toast. Once this happens, fermentation is under way. Slowly add the rest of the yeast starter to the must.

10 Once fermentation begins, it is important not to aerate the liquid anymore; so do not mix or move it. The must will form a natural barrier on the surface that protects the liquid from oxygen. On the other hand, gases from the fermentation will bubble through the must and should be allowed to escape. For this reason, it is important not to seal the plastic bucket. After three days of fermentation, cover the bucket with a few layers of paper towels only. If you begin to see fruit flies before that point, cover the bucket with paper towels immediately (making sure no flies are trapped inside).

11 This primary fermentation will last from 1 to 3 weeks. Temperature is a key component in the process. During this step, the must should be kept between 65 and 70°F because yeast requires a warm environment in order to do its work. Watch it, and listen for bubbles.

12 When the bubbling and foaming subside, strain and clear the wine by pouring it through a freshly laundered dish towel into the glass jugs. After straining, move the jugs to a cooler place and do not disturb. The ambient temperature should now be between 55 and 60°F. Air contact needs to be eliminated as much as possible. Cap off the jugs, preferably with a water-lock stopper that allows any late fermentation exhalation to escape, but will not let new air in. Old corks, loosely fitted, will also do the trick, as will cotton stoppers. Let the wine rest there for 2 to 3 weeks.

1 3 During this time, debris should fall to the bottom of the jugs. If the
wine doesn't clear by the end of 3 weeks, take cleaned eggshells
that have been dried in the oven and break them into your wine. Debris will
adhere to the shells, helping to clear the wine. Once the debris is sepa-
rated from the juice, siphoning from one container to another is best, to
avoid disturbing debris. Siphon off the top into another clean glass container, being sure not to pick up any of the debris or eggshells that have
settled. You will have to suck a little juice through your hose to get the
juices flowing. If you do not have a plastic siphon hose, you can also carefully pour the liquid at the top out into new containers using a funnel. Discard the debris.

14 When the wine is clear enough for your liking (this may take up to 3 months), it is ready for bottling. We have seen everything from plastic screw-top soda bottles to traditional corked glass bottles used for wine storage. We find that glass screw-top bottles (the type that San Pellegrino and other fancy spring waters come in) are a nice compromise between convenience and tradition. Clean them with plenty of dish soap and submerge the bottles in boiling water for at least 15 minutes to sterilize them. The bottles *must* be thoroughly sterilized before wine touches them. Once the bottles are cool, pour in the new wine, leaving 1 inch of space from the top. Screw the caps on as tightly as possible, and store in a cool, dry, dark place. The wine will be best one year from bottling, but it is ready to consume once it clears.

14a

14b

chapter 17

Cupboard **Keg**

*_Brew beer from scratch_

How do you make **something** (beer) **out of nothing** (stale rye bread, peppermint, yeast, and raisins)? Easy. Brewmaster **Ryan Horan** showed us how.

Kvass is a lightly alcoholic beverage, similar to some of the lambic-style beers of Belgium. It originated in Russia, where it is still a popular summer drink. Natural sugars found in almost any kind of fruit can be used in the brew, but for the sake of thrift, we used dried apricots and raisins. You can use raspberries, apples, and even rhubarb (stalks, not leaves, which are poisonous) when it's in season.

Equipment

Plastic or glass bowl

2 medium-size pots

Cheesecloth or freshly laundered dish towel

Medium-size glass jug

Large, empty beer bottles or wine bottles

Corks

Ingredients

1 lb loaf of stale rustic rye bread

4 quarts hot water

2 (¼ oz) packages Fleischmann's Active Dry Yeast

½ cup sugar

5 tbsp warm water

1 bunch fresh mint leaves

Raisins or other fruit

Strip of lemon zest

1 In one medium-size pot, soak the whole loaf of bread in 4 quarts of hot water. Let rest for 4 hours; then mush it up with your (very clean) hands.

2 Strain the water through several layers of cheesecloth or a kitchen towel into a second medium-size pot. Be sure to squeeze all of the water out of

the bread. Stir together the two packages of yeast, ½ cup of sugar, and 5 tablespoons of warm water in a plastic or glass bowl. Let this mixture sit for several minutes before pouring it into the brew. Add a large bunch of mint leaves (but save one leaf). Leave the brew in a warm place overnight (8 to 10 hours).

3 Clean your jug with dish soap and sterilize it by submerging it in boiling water for at least 15 minutes.

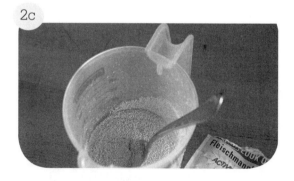

4 After its night of resting, strain the mixture once more into the sterilized jug. Add a few raisins (or pieces of other fruit), the remaining leaf of mint, and a strip of lemon zest. Tightly seal the jug with a cork and refrigerate for 3 days.

5 When the raisins bob to the surface, strain the liquid once more into a clean pot. Thoroughly clean the empty wine or beer bottles and sterilize them. When the bottles are dry and cool, funnel the liquid into them. The kvass is now ready for drinking. *Na zdrovia!*

4a

4b

5

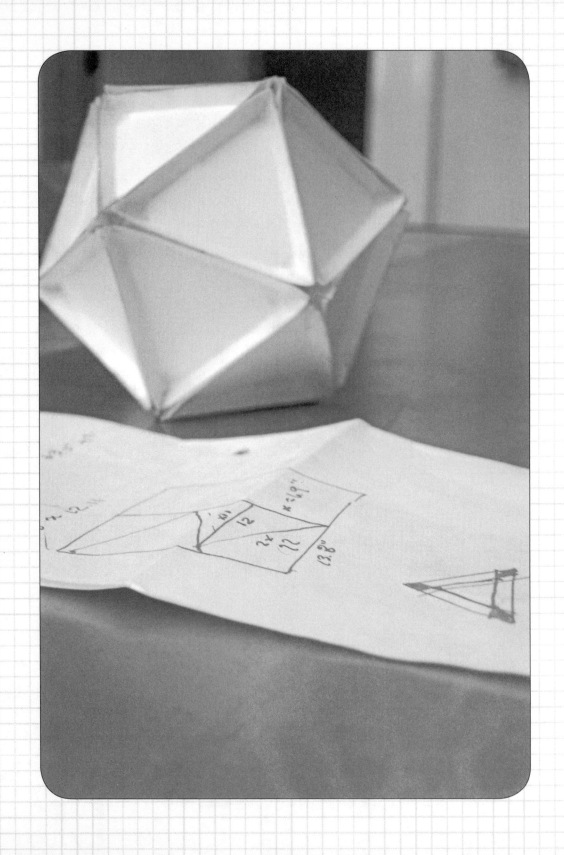

chapter 18

Pie in the Sky

* *Build a modular pecan pie*

Pie baking and gadget making can both be **competitive sports**, and when the two activities are combined, things can get a little **out of control**. According to the *Guinness Book of World Records*, the largest pumpkin pie measured more than 12 feet across and the heaviest cherry pie weighed nearly 40,000 pounds. But no one has yet gone on record for making a **three-dimensional pecan pie**—until now. The **Turkey Tek Collective** takes the cake.

The beautifully symmetric regular icosahedron is one of the five Platonic solids, or convex regular polyhedrons. It consists of 20 triangular faces, 5 of which meet at each of 12 vertices. Make a pecan pie for each triangular face, build an icosahedron with magnets, and all Hungry Geometers in the room will have pentagonal stars in their eyes.

It's hard to describe the excitement of wrestling with incredibly small and powerful magnets that are covered with a combination of pie filling and rapidly hardening epoxy. Let's just say we wouldn't spend the night before Thanksgiving any other way.

Equipment

Pie pans

2 (24×48 inch) sheets of 24-gauge steel

Power shear or hand shear

Sheet-metal bending brake or pliers

60 "POP" rivets

Rivet gun

Aluminum foil

Epoxy adhesive

180 small, flat, circular, super-strength magnets (see Appendix for where to purchase them, page 192)

Drill

Pies

Medium-size bowl

Large bowl

Pastry cutter or two knives

Rolling pin

Plastic wrap (if needed)

Ingredients

Crust

30 cups flour

10 tsp salt

20 sticks butter

Bucket of ice water

Filling

20 cups light corn syrup

60 eggs

20 cups sugar

5 sticks butter

20 tsp vanilla extract

40 cups pecans

1½ tbsp flour

1 Draw a template for 10 equilateral triangles on each sheet of steel. Each triangle should have an altitude (distance from one triangle point to the middle of the opposite side) of 12 inches. Cut out each triangle with the shears. Once your 20 triangles are cut, on each triangle mark three lines inset 1½ inches from the edges. These will be where you fold the edges up to become the sides of each triangular pan.

1a

1b

2   Next, cut off the corner triangles formed by your lines and cut a slit at each corner so that you can fold up the side walls. Using the sheet-metal bending brake (or pliers and patience), fold up the triangular tab at each corner; then fold up each side of the pan to form an approximately 120° angle. Take extra care to shape the metal accurately so that the pans will fit together when assembled. Drill a hole through each corner tab and secure with a rivet. Set aside.

3   Combine 1½ cups flour and ½ tsp salt in a medium-size bowl. Cut in 1 stick of butter (with a pastry cutter or two knives) into pea-size chunks and sprinkle in enough ice water—a tablespoon at a time—until the dough holds together to form a ball. Roll the dough out to a thickness of half a centimeter. (If it does not hold together, wrap it in plastic wrap and place it in the freezer for a ½ hour.) Gently lift the rolled dough into a pie pan. (No need to grease the pan. We don't want the pie to fall out of the pan when inverted.) In a large bowl, combine 1 cup corn syrup, 3 eggs, 1 cup sugar, 2 tablespoons melted butter, 1 teaspoon vanilla, and 2 cups pecans. Mix thoroughly and pour into the crust.

4   Repeat step three 19 times.

5   Line the bottom of your oven with aluminum foil; then heat the oven to 350°F. Fit as many pie pans into the oven as you can. Bake each batch of pies for 1 to 1½ hours, until the pies are firm in the center and the crust is golden brown.

6   Remove each batch from the oven and let them cool completely on a rack. When the pans are cool, carefully epoxy three magnets to each edge of each pan, being careful not to get epoxy on any of the pies. Space the magnets along each edge so that they don't interfere with those on the neighboring pans. It's important to put magnets on after baking, as oven heat can cause demagnetization.

2b: A planar pattern for a pie-cosahedron.

2a

2b

2c

2d

5a

5b

6a

6b

7 Once the epoxy is dry, carefully assemble the pieces and feast your eyes on the resulting *pie*-cosahedron. When it comes time to serve your guests, the disparate pies will easily disassemble.

Heart Spoon

FASHION A STETHOSCOPE OUT OF A MEASURING SPOON

It's good to keep a stethoscope around—you never know when you're going to need to take the pulse of another tinkerer lying on the kitchen floor. Stethoscopes can also be used for cracking safes; they amplify the sounds that the springs and pins make inside combination locks, revealing the sequence of turns that will unlock the loot. They are made with a diaphragm—in this case a measuring spoon—that vibrates from the sound waves created inside the body (or lock), allowing you to auscultate, or listen, to internal functions. Pressure builds inside the dome, sending the waves up the hollow tubing attached to the other side and into the ear. Thump-thump, thump-thump, thump . . .

Cut off the handle of a **cheap, lightweight metal measuring spoon**, leaving only the circular metal portion of the spoon. If you are making a one-ear (monaural) stethoscope, **drill** one hole in the back of the spoon slightly smaller than the diameter of the tube. If you want to make the two-eared (binaural) model, drill two holes to accommodate a second tube. Into the hole(s) plug one end of a 16-inch-long, ⁵/₈-inch-wide **vinyl or rubber tube**. Drill a hole partly into a **foam earplug** and attach it to the opposite end of the tube. Now you have a stethoscope; so get auscultating!

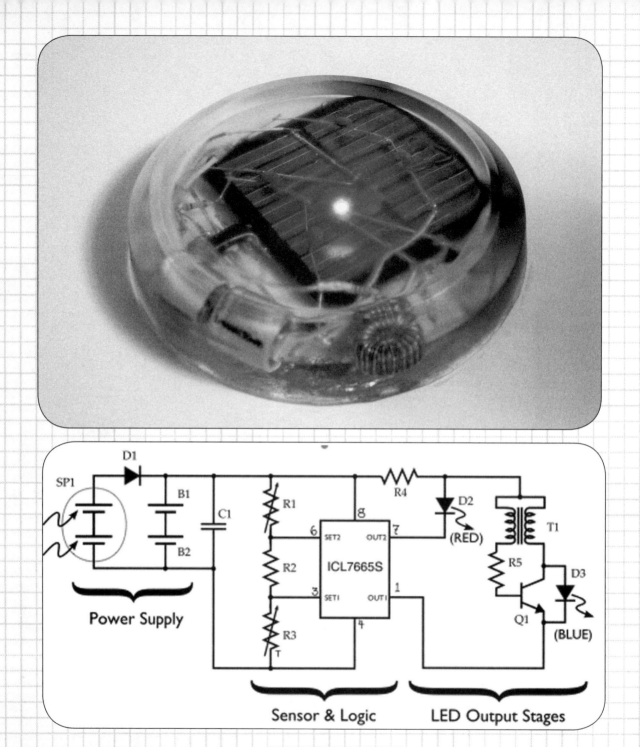

Here is the schematic diagram for the coaster circuit, which shows the
components and how they are arranged. (Fear not—it is not actually necessary to
understand this diagram in order to build a set of working coasters.)

chapter 19

Flying Coasters

* *Rig solar-powered–heat-sensitive smart coasters*

When a hot drink is placed on a "smart coaster," the coaster lights up in red. When a cold drink is placed on one, the coaster lights up in blue. Simple! Actually, *not really.* **Windell Oskay** and **Lenore Edman** have invented a gadget worthy of the **Design Hall of Fame**. Behold. . . .

The smart coaster circuitry is hermetically sealed in polyester casting resin, making it waterproof and easy to clean. Power is provided by a small solar panel—the type used for solar garden lights. An hour of direct sunlight provides enough energy, stored in a tiny two-cell NiMH battery, to run the red LED for several hours. Because the battery voltage is not high enough to directly drive the blue LED (see page 9 for LED Love), a "Joule thief" circuit (a simple, compact switching power supply) is used to produce the necessary voltage. Although the circuit continuously monitors the temperature, the standby power consumption is extremely low and will not drain the battery for more than a year.

The instructions given here are designed to enable anyone who is handy with a soldering iron to make their own set of smart coasters. Toward that end, parts of the project are presented to appeal to DIYers of different levels. While the circuit diagram and theory of the circuit are fascinating, you can also flip past those parts and build your coasters by following the steps at hand. There are two major parts to building smart coasters: building the circuit and encasing it in plastic. Coasters usually come in sets, and you'll probably want to build and test several coaster circuits before moving on to the casting stage.

You can use this circuit in many other ways. Imagine a large, transparent serving platter that has several complete circuits embedded in it or an array of solar garden lights or pool lights. . . . The sky is the limit. The circuit is yours to hack.

Equipment

For the circuit

A soldering iron

Solder

A "helping hands" soldering jig

Small pliers

Wire stripper

Fine heat shrink tubing

Small flathead screwdriver, for calibrating the circuit

For casting

Clear polyester casting resin

MEKP catalyst

Disposable food-storage containers, for molds

Disposable cups and spoons, for mixing

Clear gloss spray finish, for overcoating castings

Solvents, such as rubbing alcohol or acetone, for cleanup

A kitchen scale, for measuring casting resin

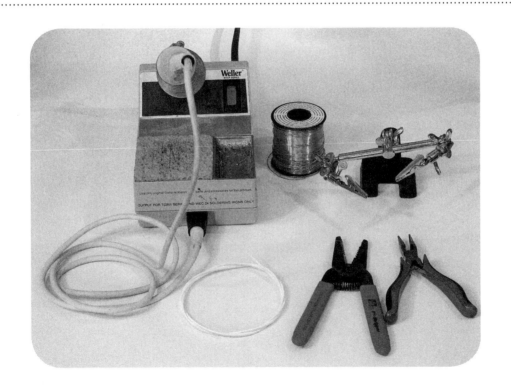

Ingredients

Note: These are the parts needed to make one coaster. To build ten coasters, you'll need ten times as many of each component. For convenience, Digi-Key part numbers are given where possible; visit their Web site (www.digikey.com) to order them. You may be able to find some of these components elsewhere at a lower cost.

Circuit Diagram Designation	Count	Description	Part No./Source
SP1	1	Solar panel, 3.5 V, 30 mA	(see Note 1 in the Appendix)
B1,B2	2	¼ AAA NiMH batteries	Batteries Wholesale
C1	1	~50 nF ceramic capacitor	399-4189-ND
D1	1	1N914 diode	1N914CT-ND
D2	1	Red LED (see Note 2)	67-1610-ND
D3	1	Blue LED (see Note 2)	516-1360-ND
ICL7665S	1	Dual voltage detector IC	ICL7665SCPAZ-ND
R1	1	1 megohm 25-turn trimmer	490-2934-ND
R2	1	150k-ohm resistor (see Note 3)	150KXBK-ND
R3	1	470k-ohm thermistor NTC	495-2162-ND
R4	1	100-ohm resistor	P100BACT-ND
R5	1	1k-ohm resistor	P1.0KBACT-ND
Q1	1	2N3904 transistor	2N3904FS-ND
T1	1	Bifilar toroidal transformer	M8670-ND
-	-	Insulated copper wire	(see Note 4 in the Appendix)

*
*See the Appendix for the notes on the following instructions,
circuit design theory, and more ordering information for parts.*

part one:
Building the Smart Coaster Circuit

1 To make the Joule thief circuit, use the blue LED, the bifilar toroidal transformer, the transistor, and the 1k resistor. With the rounded end of the blue LED facing away from you, orient the flat edge of the plastic package up, and bend the two leads to the right, until they are flush against the package. After you've done this, turn the LED over and set it on a table, adjusting it so that the rounded end of the package points straight up.

2 Locate the 2N3904 transistor and notice that the plastic package has a round side and a flat side with writing on it. Bend the two outer leads horizontally outward as shown.

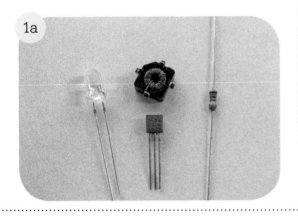
1a

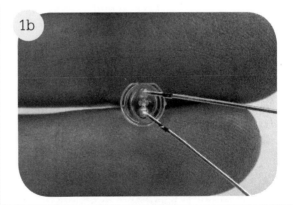
1b

1c

2

3 Now connect the transistor to the LED: Start with the blue LED pointing up as before. Turn the transistor so that the flat side with the writing faces down, and set it, as shown, across the leads of the LED. (The easiest way to do this is to use the "helping hands" soldering jig.) Gently solder the two intersections, and then trim away the excess wire from the solder joint.

4 Slip a small piece of heat shrink tubing over the middle lead of the transistor and bend that lead to the side as shown. (The tubing insulates the middle lead from the one that it crosses.) Place the 1k resistor on the middle lead, as close to the body of the resistor as possible. You can make a little loop in the resistor lead to help hold things in place before soldering. Gently solder the intersection of the transistor and resistor, and then trim away the excess wire from the solder joint.

5 Cut approximately 4 inches of copper wire (preferably with blue insulation), strip one end, and solder it to the lead of the transistor that is on the same side as the resistor.

6 Now begin to attach the bifilar-wound transformer. The transformer has four terminals or wire leads, and we want to attach two of them together. Attach two terminals (or leads) that are opposite each other (either pair will work): Cut a piece of copper wire (preferably with red insulation) about 5 inches long, and strip half an inch from one end. Lay that stripped end across both the coil and the terminals, and solder it to two terminals opposite each other as shown.

7 Attach a 2-inch length of (preferably blue) copper wire to one of the two unused terminals (or leads) of the transformer.

8 Solder the other unused terminal of the transformer to the unused end of the resistor; but for the sake of compactness, first trim the lead of the resistor to the minimum size that you need to fit the components together neatly.

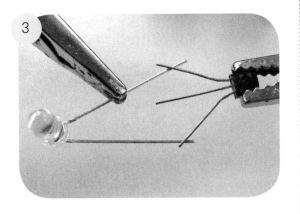

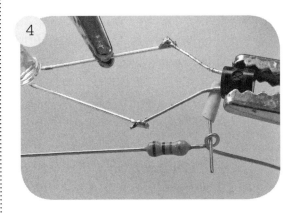

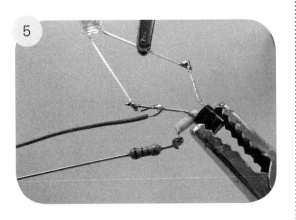

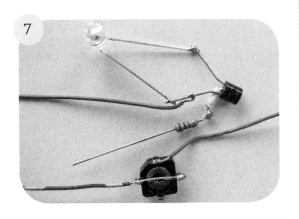

9 Connect the other end of the short wire (2-inch) to the far lead of the transistor as shown. Gently solder the joint, and then trim away the protruding wire ends.

10 Your new Joule thief circuit is complete. Testing it is a critically important step; you MUST verify that it works before integrating it into the rest of the circuit. You should have two long wires coming off the circuit; one (red) is connected to the transformer and the other (blue) is connected to the transistor. To test it, strip the ends of those two wires and press them against the ends of a 1.5-V battery cell; red to the positive end of the battery, blue to the negative end. If it lights up, your circuit is working. If not, back up and make sure that all of your connections are where they are supposed to be.

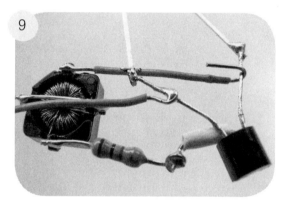

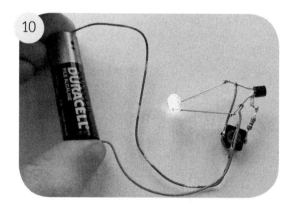

[JOULE **THIEVES**]

Congratulations—you've built a working Joule thief! What can it do?

The Joule thief is a simple gadget for switching power supply. It takes the 1.5-V signal from a battery cell and boosts it above the approximately 3.5 V necessary to drive the blue LED. It works by pulsing the output from the battery, creating a series of short pulses that individually exceed 3.5-V. (The oscillation frequency is typically around 50 kHz; you will not see it blinking.) This is an extremely handy circuit to use in a variety of circumstances, since it can raise the low voltage from a battery high enough to do something useful. This circuit is called a Joule thief because it allows you to steal every last bit of energy (a Joule being a unit of energy) stored in a battery; you can often use it to drive an LED with an apparently dead battery cell.

part two:
Preparing the Chip

11 The chip is an ICL7665S voltage detector, which comes in an eight-pin plastic package. Using a pair of pliers, carefully bend the leads so that they all lie flat in one plane. We'll be referring to the pins by number, so let's get the numbering scheme straight by making a graphic. The pins of a chip like this are numbered counterclockwise, starting at the position indicated on this graphic by a little circle. You aren't going to need pins 2 and 5 in the circuit. In order to keep things neat, we recommend removing those two pins. To do so, gently grab the pin with a pair of fine-tipped pliers and gently bend it back and forth several times until the metal weakens and breaks off. Do not yank, tear, or clip the pins off; if you do, you risk damaging the silicon chip itself.

12 Attach the capacitor to the chip, connecting pins 4 and 8: Rest the body of the capacitor on top of the chip, bend its two leads to touch pins 4 and 8 of the chip, and gently solder them in place.

13 Attach resistor R2: (In the pictures we used a 100-kilo-ohm resistor, but 150k is usually a better choice; see Note 3 in the Appendix [page 194] for an explanation of this point.) Place the body of the resistor against one of the leads of the capacitor so that there can be no physical contact

[SOLDERING **SMARTS**]

You must be careful not to overheat chips when soldering to them; if you aren't able to make a good connection in the first few seconds of soldering, wait a few minutes for the chip to cool before trying again.

between the leads of the capacitor and the leads of the resistor. Using pliers, wrap the leads of the resistor around pins 3 and 6 to hold them in place before soldering. Gently solder the resistor to those two pins, and trim away the extraneous ends of the resistor leads.

14 Solder a 3-inch segment of (preferably red) wire to pin 7. That completes the preparation of the chip.

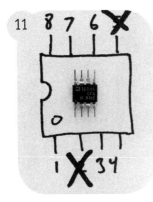

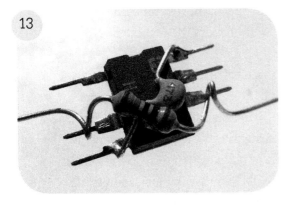

part three:
Preparing the Power Supply

In these steps, you'll need the solar panel, the two battery cells, the 1N914 diode (D1), and a couple of pieces of plain copper wire several inches long, which you can obtain by stripping the insulation from the insulated wire.

15 The back of the solar panel has some solder connection terminals, which should be clearly labeled positive and negative. To the positive terminal, solder the 1N914 diode, with the black band pointing away from the solar panel. Try to minimize the distance from the solar panel to the body of the diode. Solder several inches of plain copper wire to the negative terminal. The back side of the solar panel should now look something like this:

16 Connect the batteries: These NiMH battery cells have a solder tab (with a hole) on the positive end and a bare terminal on the negative end. In order to solder something to the negative terminal, you first need to "wet" the solder to that terminal: Melt a small blob of solder onto each negative terminal, and make sure that it is genuinely sticking to the surface. This can be a little tricky and can take a little time. If you don't succeed in about fifteen seconds, let the battery cool for about fifteen minutes before trying again. In any case, the battery may be too hot to touch for some time after soldering to it. Once you've wet the negative terminals, solder the positive tab of one cell to the negative terminal of the other. Bend the tab that connects the two battery cells so that the two cells line up together. Finally, solder another piece of uninsulated copper wire to the remaining negative terminal.

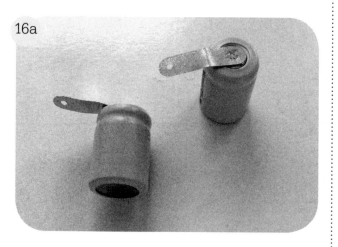

16a

16b

16c

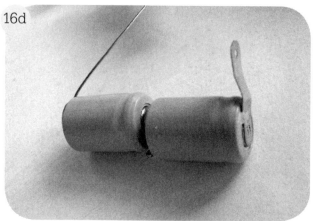

16d

part four:
Put it All Together

17 Turn the solar panel right-side up. Take the battery assembly and slide the hole in the solder tab of the remaining positive terminal over the exposed end of the 1N914 diode. Place the prepared chip assembly between the terminals of the solar panel as shown:

18 Bend the uninsulated copper wire onto the solar cell and solder it to pin 4 of the chip. Bend the exposed end of the 1N914 diode and solder it to pin 8 of the chip. As usual, clip away excess wire after soldering.

17

18

19 Add the Joule thief circuit: Place the blue LED atop the solar panel as shown. It should lie flat, with the round end of the LED pointing straight up. To attach the blue wire from the Joule thief (the one attached to the transistor) to pin 1 of the chip, first trim it to length and then strip the exposed end before soldering it in place. If you've got some handy, use a piece of Scotch tape to help hold the LED in place while you do the next few steps; just remember to remove it before potting the circuits.

20 Prepare the trimmer: The trimmer has three staggered leads. Carefully clip off the first lead. Bend the third nearly flush against the package as shown.

21 Install the trimmer: Connect the two remaining leads from the trimmer to the leads coming off pins 6 and 8 of the chip.

22 Finish connecting the battery: Move the battery to lie next to the chip as shown. Take the uninsulated copper wire that comes from the negative terminal of the battery and solder it around the uninsulated copper wire that attaches to pin 4 of the chip, minimizing the amount of slack wire and trimming the excess after soldering. Now solder the tab on the positive end of the battery to the end of the 1N914 diode that is attached to pin 8 of the chip. At this point, the circuit is live in the sense that the solar cell can charge the battery and the battery provides power to run the chip. That's okay—you can continue to build the circuit.

23 Prepare the red LED and thermistor: Bend the leads of the red LED (D2) in the same way as you did in step 1. The orientation is important. Now bend the leads of the thermistor (R3) so that it sits up at about the same height as the top of the LED. (You can bend the leads of the thermistor in either direction.)

19a

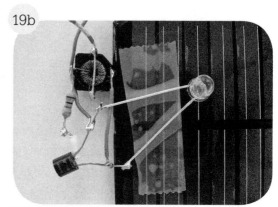

19b

20a

20b

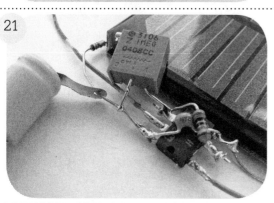

21

22

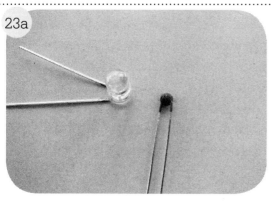

23a

23b

24 Attach the thermistor: Set the thermistor on the solar panel, somewhat close to the blue LED. It should stand approximately the same height as the LED. The two leads attach to pins 3 and 4 of the chip. It's straightforward to solder one of the leads to the bare copper wires that connect to pin 4 (and the negative terminals of the battery and solar cell). To attach the other lead to pin 3, you may need to bend it a little so that it does not touch the wires connected to pin 4. Once you've completed this, you may find it helpful to hold the thermistor in place with a small piece of Scotch tape.

25 Solder one end of R4, the 100-ohm resistor, to the lead coming off the black-banded end of the 1N914 diode. Bend it in place as shown, so that it lies parallel to the diode.

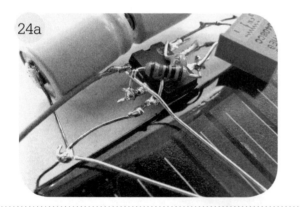

24a

24b

25

26 There is still one loose (red) wire coming off the Joule thief end of the circuit. It's the one connected to two terminals of the bifilar-wound transformer. Snake it under the batteries and over to meet the end of the 100-ohm resistor that you just added. Strip the end of the wire and solder it to the end of the resistor that isn't touching the diode. While you should trim the excess copper wire after soldering, don't trim the resistor lead just yet. (Since this completes the circuit for the Joule thief, it is possible that your blue LED will light up during this step.)

27 Add the red LED: Now place the red LED, with its leads bent (like those of the blue LED) atop the solar panel, close to the thermistor and the blue LED. The leads of the red LED should face to the right. Solder the uppermost lead to the end of the 100-ohm resistor.

26

27

28

There is only one connection remaining. The only loose wire that you should have now is connected to pin 7 of the chip, and it needs to get connected to the other lead of the red LED, the lowermost one. Cut the wire to length, strip the end, and solder it in place. (Since this completes the circuit for the red LED, it is possible that it will light up during this step.)

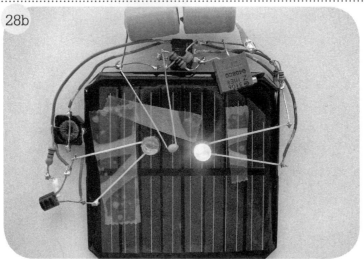

part five:
Testing and Calibrating the Circuit

29 Using a small flathead screwdriver, carefully turn the trimmer's adjustment screw through its 25-turn range. (When you get to the edge of the range, it will make a quiet clicking noise once per turn, letting you know that you should turn it the other way.) If everything is working, the red LED should light up at one end of the range, the blue LED at the other end, with a narrow band in the middle where neither LED is on. Turn the trimmer back to the middle place where both LEDs are off, and leave it set there.

To test your circuit's thermosensitivity, pinch the thermistor between your fingers; the red LED should light up within a minute or so. To see the behavior in a cold environment, you can put the circuit in the refrigerator for a minute and see if the blue light turns on. Once the circuit is tested and working, the only adjustable parameter is the position of the trimmer. It controls the temperature set point, which is the temperature at which both LEDs are off. Since room temperatures will always vary over some range, there is a small range of temperature near the set point where both LEDs remain off. That way, the batteries won't get drained every time your heater or air-conditioning comes on.

For proper operation, the set point must be calibrated to reflect the actual room temperature that the coasters will be stored in. To do this, first leave the circuit alone for an hour in a place that is at your normal room temperature—not in a sunbeam, on top of (or in) the fridge, or right next to the air-conditioning vent. Then, carefully turn the trimmer right to the middle of the band where the LED turns off. If you accidentally touch the thermistor during the process, let the circuit sit for a while again before trying to tune it to the middle of the band.

29a

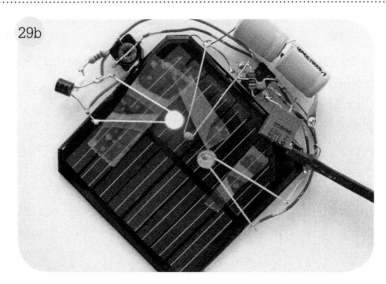

29b

29c

If the LEDs behave in any other manner than what's been described, it is possible that you have made a wiring error; go back and check all of your connections. If neither LED turns on in the full range of the trimmer, it is possible that your battery needs charging, so set the circuit in sunlight for an hour and then try again.

CAUTION: Methyl ethyl *ketone peroxide—aka catalyst—is combustible, can cause severe burns, and may be fatal if swallowed. So don't work with resin or catalyst around food. Don't pour excess resin into the sink; it will clog the drain. And please work in a well-ventilated room. Thank you.*

part six:

Potting the Circuit

Now that you've built and tested a few smart coaster circuits, it's time to turn them into coasters, which means casting them in plastic.

Casting is mostly a matter of following the manufacturer's instructions, and mistakes can easily be made. Use too little catalyst and your goo doesn't solidify; use too much and differential shrinkage of the resin will shatter your finished casting. Since things can go wrong in the casting process, you may want to test your casting process before putting your nice circuit in plastic.

Read these instructions and the instructions for the casting resin. Make sure that you have everything that you need in advance, because there's no going back once the resin is poured. If you have used Scotch tape to hold down the LEDs on your circuits, remove it now, before going on to the next step.

3 0 In a disposable plastic container, working with disposable plastic tools, measure out several ounces of resin; enough to fill several molds about ⅜ inch deep. Follow the mixing instructions on the casting resin for making a two-layer (multilayer) casting, and add the required amount of catalyst to the resin. Stir the resin and catalyst very well to mix them, for about 1 minute. Try to avoid introducing air bubbles by mixing in a smooth consistent manner. The resin will change color slightly as the catalyst is stirred in, as a subtle indication of the reaction that is taking place. DO NOT try to mix up resin for more than three or four castings at a time, since the resin may start to harden before you have time to distribute it.

31 Pour the mixed resin into the molds, up to a depth of about ¼ to ⅜ inch.

32 Quickly, while the resin is still wet, place a circuit, upside down, in the wet resin and push it down gently. You want the LEDs and thermistor to almost rest on the bottom, so that when the casting is done, they are covered by a thin layer of plastic. If you can see any bubbles trapped beneath the circuit, do your best to remove them by gently tapping the molds before the plastic sets.

33 Once the first layer begins to gel, in about 5 minutes, mix up an additional batch of casting resin. This next layer will be the top layer of the casting, so adjust the amount of catalyst per the resin package's instructions.

34 Pour the casting resin on top of the circuits to form the second layer of the casting. You want to make the thinnest possible casting that (1) completely covers the circuit elements and (2) is flat on the top surface.

35 After the casting, wait 24 hours before unmolding the coasters. If they have not cured correctly and are excessively tacky, one trick that can help is to place them on a cookie sheet in a conventional oven at 200°F (no higher) for several hours. Very minor remaining tack can be sealed in place by using the spray finish.

36 Charge the coasters and use them. Give them their initial charge by leaving them outside on a sunny day for a few hours. When not in use, store them indoors, away from hot or cold areas, so as to preserve the battery life. Recharge them as needed.

BRIGIT BINNS (Mallow Ammo, page 111) is the author or coauthor of nineteen cookbooks, most recently *The Relaxed Kitchen* (St. Martins, 2007). She is responsible for the culinary education of her niece, Hungry Scientist coauthor Lily Binns (and Lily is responsible for Brigit's culinary miseducation).

JENNIFER BOTTO (Basement Bacchanalia, page 129) is a twenty-five-year-old non-native Bostonian, eager to return to her farming roots. She began studying wine making a few years ago (her first attempt produced 4 gallons of fine vinegar), especially the history of and recipes for folk wines; that is, traditional wines made from fruits, vegetables, flowers, and herbs. She hopes to someday own a vineyard and farm of her own.

CHRISTIAN BROOKFIELD (Magnet Madness, page 37; Bar None, page 39) enjoys going to the Exploratorium in San Francisco. He is learning to play both the trombone and the guitar but not simultaneously. A contributing writer at www.evilmadscientist.com, he also likes marine biology and electronics. He rides his bike to school and, like many of his classmates, complains about his homework. His favorite color is yellow; not yellow like a lemon, but yellow like a school bus.

LENORE EDMAN (Dip 'n' Dots, page 31; Bar None, page 39; Edible Origami, page 93; Lego Trebuchet, page 117; Flying Coasters, page 155), a veteran bike commuter, used to live in Portland, Oregon, where her son Christian got to ride in her bike's sidecar. Abandoning wet for warm, she moved to Austin, Texas, where she designed and sewed her own wedding dress. Later, as a regular of the Boulder, Colorado, weekly cruiser bike ride, she overhauled a mid-century Hawthorne ladies bicycle (named Stella), which she has crashed only once—and it wasn't her fault. Since moving to Sunnyvale, California, she has helped to popularize edible origami and has learned to make some wicked curries. She and Windell Oskay are the founders of www.evilmadscientist. com, a Web site that legitimizes some of her otherwise nearly inexplicable projects.

DAN GOLDWATER (Delectable Diodes, page 7) is an electrical engineer who eats LEDs for breakfast. He has published several tutorials and DIY projects using LEDs (both practical and whimsical) at www.instructables.com, which, in addition to Oakland-based Squid Labs, he helped to cofound. Dan is optimistic that LEDs will soon replace the ubiquitous but dangerous forms of legacy lighting.

KATE GUSMANO (Pumpkin Pin-Up, page 15) hails from Sparkill, New York. Growing up, Kate enjoyed organizing canned goods, playing with her golden retrievers, Hagel and Roman, and taking camping trips in the family VW pop-top. As a grown "woman," she is equally at home living in the woods, studying the behaviors of her avian neighbors, or roaming the world with an assortment of unconventional cameras. When she's not hollowing out her food,

Ms. Gusmano enjoys baking fruity vegan cookies, exfoliating, and training to be a *luchadora*. She recommends wearing goggles more often than not.

TURKEY TEK COLLECTIVE (Pie in the Sky, page 145; Gravy Train, page 101; I Scream for Cryogenic Ice Cream, page 55) is a group of nerds who've been getting together for the past six years to celebrate the Thanksgiving holiday. Thanksgiving is a chance for them to come together as a large, extended family of friends for four days of fun and games. Folks fly or drive from across the country to participate. Each year, there is a natural tendency to outdo whatever went down the year before—the sort of group mentality where you're really pushing one another to stretch the limits. This naturally leads to an unstable positive feedback loop of one-upsmanship, or mutual assured destruction, through increasing amounts of food presented in ever more absurd fashion.

RYAN HORAN (Hot Box, page 85; Living Loaf, page 121; Basement Bacchanalia, page 129; Cupboard Keg, page 139), a native of Cincinnati, is interested in applying traditional food production methods to a modern setting. He has worked on a Tennessee river barge, as a zookeeper, on reality TV, as a mover, as a rowing coach, and as a copier salesman.

WINDELL OSKAY (Dip 'n' Dots, page 31; Bar None, page 39; Flying Coasters, page 155) owns only one slide rule but plans to acquire more. A published playwright, award-winning cartoonist, and obscenely creative amateur chef (several people have described food that he has prepared in terms more positive than "edible"), he has been variously employed as a quantum mechanic, photographer, and (atomic) clock maker. Some of the other things that he somehow ended up designing and building include an interactive dining table, a carbon-fiber electric guitar, hard-drive wind chimes, model hovercrafts, nixie tube clocks, and magnetohydrodynamic-powered boats. He and Lenore Edman founded the Web site called www.evilmadscientist.com in order to begin documenting this craziness.

JENNA PHILLIPS (Edible Undies, page 1) often finds herself engaged in an irreverent mixture of science and fashion. Formerly a lingerie designer in Paris, she presently runs her own high-tech shoe company called Formula Magic. Jenna holds degrees in history and literature from the University of California, Berkeley, and Oxford University, where she had a penchant for presenting lingerie runway shows in the world's oldest debating chamber. Jenna is a Hungry Scientist because during her unflagging research, she herself often forgets to eat!

RICK UNGER (Toasty Paws, page 50) considers himself a somewhat handy guy. He likes to create things, frequently as gifts for friends who inspire him. A photographer, sculptor, and musician, Rick often collaborates with other artists to really get his creative juices flowing.

LEE VON KRAUS (Tupperware Party, page 71) was born in New Jersey and grew up in Boston, Massachusetts. At the moment, he is thoroughly enjoying a neuro-robotics doctorate program at SUNY Downstate Medical Center in New York City and pursuing all that the city has to offer.

MICHAEL ZBYSZYÑSKI (Intergalactose Scream, page 75) is a composer, sound artist, performer, and teacher in the field of contemporary electroacoustic music. Currently, he is Assistant Director of Music Composition and Pedagogy at UC Berkeley's Center for New Music and Audio Technologies.

David Picard: photographs, pages xiv, 5.

Louis Camille Maillard Organization (public domain): photograph, page 2.

U.S. Patent Office (public domain): illustration, page 3, 9 (right).

Dan Goldwater: photographs, pages 6, 11(upper right; bottom), 12.

Wikipedia (public domain): photograph, page 9 (left).

Patrick Buckley: photographs, pages 11 (upper left), 22, 28, 29, 60, 63–66, 68, (sidebar top), 91, 110, 113, 115–16, 118, 136–37, 141, 151, 152. Drawing, page 95.

Kate Gusmano: photographs, pages 14, 19, 21.

Windell H. Oskay and Lenore Edman: photographs, pages 30, 33, 35–36, 38, 42–43, 45, 47–49, 92, 97–98, 118, 154, 157, 159, 161–62, 165–69, 171–74, 176, 178–79, 181.

Christian Brookfield: photographs, page 37.

Rick Unger: photographs, pages 50–52.

Charless Fowlkes: photographs, pages 54, 59, 100, 103, 105–6, 144, 147, 149–50.

Lily Binns: photograph, page 67.

Emilie Lincoln: photographs, page 68 (sidebar, bottom).

Lee Von Kraus: photographs, pages 70, 73.

Michael F. Zbyszyñski: photographs, pages 74, 77, 79, 81–82.

Ryan Horan: photographs, pages 84, 87–89, 120, 125, 127, 138, 142,(bottom right).

Jennifer Botto: photographs, pages 128, 131, 133, 135.

Martin Tolliver: photographs, page 142(top, lower left).

Contributor photos courtesy of the contributors, pages 183–87.

Hungry Scientist
Kitchen Survival

Edible Undies, page 1

For more information on candy making and supplies, see www.baking911
.com.

Delectable Diodes, page 7

For more information on LED voltage and wiring instructions, see the Electronics Club Web site, at www.kpsec.freeuk.com/components/led.htm.

Pumpkin Pin-Up, page 15

For photo-developer materials, see www.bhphotovideo.com.

Party Like It's 2099, page 23

For a source for *varak*, or silver leaf, see www.kalustyans.com.

Bar None, page 39

For a guide to where to find dry ice at a store near you, check out www
.dryicedirectory.com.

I Scream for Cryogenic Ice Cream, page 55

For safety equipment to use with liquid nitrogen, check out www.fishersci
.com.

Warm Bud, page 61

For more can projects, look out for this classic book on eBay: *How to Use Tin
Can Metal in Science Projects*, by Edward J. Skibness (T. S. Denison, 1960).

Hot Box, page 85

For more tips and recipes, see www.cajachinaperu.com.

Pie in the Sky (page 145)

For heavy-duty magnets, check out www.amazingmagnets.com.

Flying Coasters (page 155)

Where to Get Tools

Helping Hands:	Hobby Engineering (item #1666); see "Where to Buy Electronics," below
	Electronic Goldmine (item #G9773); see "Where to Buy Electronics," below
Small pliers:	Sears, Craftsman Professional series

We used TAP Plastics Clear-Lite casting resin (1 qt: $20.50), catalyst (½ oz: $2.45), and finishing spray (5.25 oz can: $5.25). However, you can also order equivalent items from other companies, such as Dick Blick. (See "Where to Buy Electronics," below.)

Where to Buy Electronics

All Electronics:	www.allelectronics.com
Batteries Wholesale:	www.batterieswholesale.com
Digi-Key:	www.digikey.com
Dick Blick:	www.dickblick.com
Andy's Solar Bugs:	www.solarbug.com/solar-kit.html
Tap Plastics:	www.tapplastics.com
Solarbotics:	www.solarbotics.com
The Electronic Goldmine:	www.goldmine-elec.com
American Science & Surplus:	www.sciplus.com
Hobby Engineering:	www.hobbyengineering.com
Plastecs:	www.plastecs.com
Sundance Solar:	www.store.sundancesolar.com
Sure Electronics:	www.stores.ebay.com/Sure-Electronics
Surplus Shed:	www.surplusshed.com

Note 1

Your solar panel needs to be the 3.6-V type, rated between 25 and 100 mA. (If you use a higher voltage solar panel, you must increase the value of R4 to ensure that the maximum current flowing through R4 is less than or equal to 25 mA.)

While any solar panel that meets these specifications will do the job, a common variety is that designed for solar garden lights. The typical specification of these is 3.6 V, 50 mA in a weatherproof 2.375-inch square package. Naturally, the best source is to recycle them from solar garden lights. At the time of this writing, you could also obtain the exact part from Surplus Shed, item # R3127. Thin, flexible solar panels are also now available with similar specifications, for example Sundance Solar # 700-50041-00.

The following sources have recently offered these or similar small solar panels, but may not have exactly the type specified: Sure Electronics, Plastecs, All Electronics, Solarbotics, The Electronic Goldmine, American Science & Surplus, and Hobby Engineering. See the preceding vendor list for contact information.

Note 2

The two LEDs are standard high-brightness types in T-1 ¾ (5 mm) packages. To save a few dollars and get a brighter result, get your LEDs from one of the surplus electronics shops instead.

Note 3

The choice of R2 determines the width of the temperature range at which both LEDs are off. Values in the range of 100k to 200k are reasonable. Lower values of R2 will make the coaster more sensitive by reducing the temperature range that the circuit considers to be room temperature. Thus, it will be more sensitive to the temperature of your beverage, but also more likely to turn on—and discharge the battery—because of room temperature variations. A higher value will mean that your LEDs will not light up unless there is a fairly large temperature change.

Note 4

You will need a supply of fine insulated solid-core copper wire, 22–26 gauge. Wires taken from telephone or network cable (e.g., category-5 cable) fit the description and also come in a variety of colors.

Design and Theory of the Smart Coaster Circuit

The smart coasters are hermetically sealed so that they can handle dripping condensation and cleanup under the kitchen tap. This precludes any sort of battery hatch or power connector. The power is provided by a two-cell NiMH battery, which gives an output voltage of about 2.4 V. In order to charge the battery, we use a garden-variety (pun intended) solar panel that puts out about 3.5 V. A blocking diode (D1) prevents the battery from discharging backward through the solar panel at night.

The heart of the circuit is an ICL7665S voltage detector. This micropower IC contains a voltage reference, two comparators, and two output switches. The chip is happy to run off 2.4 V, but it can be sensitive to rapid changes in voltage. Capacitor C1 has the sole purpose of protecting the chip when you first hook up the battery (it isn't actually needed once the battery is hooked up—but there's no harm in leaving it there).

The internal voltage reference of the ICL7665S is near 1.3 V. The internal comparators look at this internal reference as well as the two voltage inputs on pins 3 and 6, called SET1 and SET2. The state of the chip's outputs OUT1 and OUT2 (on pins 1 and 7) are determined by the comparators. The output OUT1 is on if $V(SET1) > 1.3$ V, and off otherwise. The second channel is just the opposite: Output OUT2 is on if $V(SET2) < 1.3$ V, and off otherwise.

What does it mean to say that OUT1 is "on"? In typical comparators, the output is a digital "high" or "low" logic signal. However, in the case of the ICL7665S, when OUT1 is "on," it actually turns on an internal switch that effectively connects OUT1 to ground. When OUT1 is "off," the connection to ground is removed. We can use this in the circuit by making a little circuit (say, an LED and a resistor) that would be a complete circuit if the other end were connected to ground, and then allowing—or not allowing—that connection to take place. In practice, the outputs connect to ground either a red LED, lighting it up, or a Joule thief circuit that drives a blue LED.

In our circuit, the inputs to SET1 and SET2 are provided by a voltage divider that contains a 470k-ohm negative temperature coefficient (NTC) thermistor. A thermistor is a highly sensitive temperature sensor (hence the name). The resistance of ours is approximately 470k-ohms at room temperature, and the resistance decreases rapidly as the temperature increases.

In the circuit diagram, there are three resistors on the left side of the chip: R1, R2, and R3. R1 is a "trimmer," a small, multiturn variable resistor. R2 is a fixed resistor that determines the width of the temperature range for which both LEDs are off, and R3 is the thermistor. The three resistors are in series with the battery and so have a total voltage of 2.4 V across them. This constitutes a voltage divider that determines the values input to SET1 and SET2. Once set up properly, the room temperature values of the two inputs are in the neighborhood of V(SET1) = 1.15 V and V(SET2) = 1.4 V. Since V(SET1) < 1.3 V and V(SET2) > 1.3 V, both output stages are off, and neither LED will be lit up. If the temperature rises, the resistance of the thermistor falls so that V(SET1) is about 1.0 V and V(SET2) is about 1.2 V. In this case, V(SET1) is still less than 1.3 V, but now V(SET2) is less than 1.3 V, so OUT2 switches on. This is our indicator for high temperature, so we can use OUT2 to turn on the red LED. If, however, the temperature falls, so that the resistance rises, we might get V(SET1) = 1.35 V and V(SET2) = 1.56 V. In this case, it is OUT1 that turns on, so we can use it to control the blue LED.

The circuit does not have a power switch; it continuously monitors changes in temperature. As such, we must work to minimize the amount of power that is used by the circuit so that we do not drain the battery too quickly. The ICL7665S draws 3–10 microamps of supply current. As long as (1) neither of the LED drive circuits is on and (2) the reverse blocking diode is on the solar panel output, the only other place that we are using current is in the voltage dividers. We choose high values for the thermistors and resistors so as to minimize the current draw. If (R1+R2+R3) is on the order of one megohm, about 2.4 microamps of current flows through the voltage divider. The total current usage of the circuit is then expected to be 5–15 microamps. The ¼ AAA NiMH cells have a capacity of 150 mAh, so a fully charged battery should not be drained for more than a year. In contrast, you can use up to 20 mA when either LED is on, so you might want to set the coasters out in the sun after you've run them for a while.

The last part of the circuit is the LED output stage. There are two parts to this. The simple part is for driving the red LED. That's just the LED itself and a current-limiting resistor. The tricky part is to drive the blue LED. Every diode has a "forward voltage" that determines how much voltage you need to put across it before current will flow. For a typical diode like the 1N914, that's close to 0.7 V. For a typical red LED, it's closer to 2 V. That's fine, because our battery is at 2.4 V. However, a typical blue LED requires 3.5 V before it will turn on. Since we don't have that in our circuit, we use a nifty building block called a Joule thief.

R4 is a current-limiting resistor that lets the LEDs run longer, but dimmer. You can "hot-rod" your coaster by using a lower value resistor here to get brighter output. However, you must be careful to measure the amount of current that the output stages draw: The maximum current sinking ability of the ICL7665S is 25 mA. If you exceed this value, you can destroy the chip.

It turns out that the coasters have a built-in low-battery indicator. The 2.4 V reference that we divide comes directly from the NiMH battery. When the battery is close to fully discharged, the voltage begins to drop, and the divided signal into V(SET2) becomes less than 1.3 V. When this happens, the red LED switches on.